# YOUR KNOWLEDGE HAS VALUE

- We will publish your bachelor's and master's thesis, essays and papers

- Your own eBook and book - sold worldwide in all relevant shops

- Earn money with each sale

Upload your text at www.GRIN.com and publish for free

**Bibliographic information published by the German National Library:**

The German National Library lists this publication in the National Bibliography; detailed bibliographic data are available on the Internet at http://dnb.dnb.de .



**Imprint:**

Copyright © 2017 GRIN Verlag, Open Publishing GmbH
Print and binding: Books on Demand GmbH, Norderstedt Germany
ISBN: 9783668550346

**This book at GRIN:**

http://www.grin.com/en/e-book/375226/an-overview-on-balancing-and-stabilization-control-of-biped-robots

Hayder Al-Shuka

# An Overview on Balancing and Stabilization Control of Biped Robots

GRIN Publishing

# An Overview on Balancing and Stabilization Control of Biped Robots

Hayder F. N. Al-Shuka[1,2]

[1]School of Control Science and Engineering, Shandong University, Jinan, China
[2]Mechanical Engineering Department, Baghdad University, Baghdad, Iraq

Table of Contents

**Abstract**

Researchers dream of developing autonomous humanoid robots which behave/walk like a human being. Biped robots, although complex, have the greatest potential for use in human-centered environments such as the home or office. Studying biped robots is also important for understanding human locomotion and improving control strategies for prosthetic and orthotic limbs. Control systems of humans walking in cluttered environments are complex, however, and may involve multiple local controllers and commands from the cerebellum. Although biped robots have been of interest over the last four decades, no unified stability/balance criterion adopted for stabilization of miscellaneous walking/running modes of biped robots has so far been available. The literature is scattered and it is difficult to construct a unified background for the balance strategies of biped motion. The zero-moment point (ZMP) criterion, however, is a conservative indicator of stabilized motion for a class of biped robots. Therefore, we offer a systematic presentation of multi-level balance controllers for stabilization and balance recovery of ZMP-based humanoid robots.

Keywords: Biped robot; Stability; Zero-moment point; Balance; Multi-level control

## 1. Introduction

Industrial robots are often rigidly attached to the ground. In contrast, mobile robots are able to move in specific environments depending on the tasks they are required to perform. Much attention has been paid to the design and control of mobile robots because they have significant characteristics such as versatile mobility and can sense and reacting to specific environments [1]. Mobile robots can be classified as legged robots, wheeled robots and tracks.

Legged robots offer significant advantages over wheeled robots and tracks in terms of workable environments, energy consumption (e.g. biped robots could be exploited to consume less energy than wheeled robots when walking on sand environments) and adaptability [2]. Although wheeled robots are simple lightweight structures, they need regular terrains for motion and lack for instance the ability to climb stairs. Tracks help to overcome this drawback, but they consume a lot of energy because of the high amount of friction between the chains and the ground. In contrast, legged robots can move in regular and irregular terrains with versatile mobility. With configuration changes, they can easily adapt to irregular environments. The small contact areas of their feet mean legged robots can be efficiently operated [2].

The biped robots, a type of two-feet legged robots, are designed to imitate human-like locomotion and perform certain tasks such as activities in danger environments, assistance to the elderly and entertainment [3-11]. The biped robot can consist of a trunk and legs with/without feet or even have an entirely human-like mechanism depending on the desired application.

Biped robots have advantages over multi-legged robots. They have considerably higher adaptability, enabling them to overcome obstacles like narrow paths or stairs with ease. This adaptability is particularly important when the robot is required to perform human-centred tasks. Because of the smaller ground contact area and fewer actuators used, their energy consumption can be lower than that of multi-legged robots. This is consistent with the assumption that two-legged animals have higher efficiency and adaptability than multi-legged animals [12]. For a historical review of legged machines, we refer readers to references [3,13].

Some challenges encountered in the design of biped robots are as below:

- Biped robots have unstable structures because of the passive joint located at the unilateral foot-ground contact [14-17].
- Because of the unilateral foot-ground contact and the varying configurations throughout the gait cycle, their mechanical description is highly nonlinear. During the single-support phase, the robot is under-actuated, turning into an over-actuated system during the double-support phase, however [18]. Consequently, the dynamic description and control laws change during transition from one phase to another [15]. It is notable that the biped robot can be fully actuated during the swing phase, with some limitations. In fact, the dynamics of the robot depend on which legs make contact with the ground. Moreover, exchange of leg support is accompanied by an impact disturbing the robot's motion [15].
- Biped robots have many degrees of freedom (DOFs). A humanoid robot may have more than 30 degrees of freedom, making its mechanical behaviour and control difficult [19-21]. To avoid this difficulty most researchers have used simple models based on approximations and assumptions. A trade-off between simplicity and accuracy is necessary.
    - Biped robots interact with different unknown environments. The surface could be elastic, sticky, soft or stiff. This requires robust algorithms for the generation of reference trajectories and control. These algorithms should be insensitive to possible disturbances and noises. Stabilization and online adaptive control schemes could solve this dilemma.

These challenges are associated with mechanics, control, electronics, artificial intelligence and human anatomy and so studying biped robots is an interdisciplinary exercise. Unified solutions are therefore difficult to develop. For further reading on difficulties encountered in biped design, we refer to references [22]. In our previous paper [23], we concentrated on discussion of the different methods used for generation of biped walking patterns and the stability/balance criteria used for biped stabilization. In this paper, we have extended discussion of the stability/balance problem to include some other criteria used for balance recovery, and we have attempted to present systematically multi-level control systems of biped robots based on ZMP criterion.

The remainder of the paper is organized as follows. The gait cycle of biped locomotion is presented in Section 2. Section 3 discusses the stability/balance criteria of the biped mechanism, and Sections 4, 5 and 6 introduce multi-level control of biped locomotion based on ZMP criterion. Section 7 concludes.

*Remark 1.* Despite numerous 3D applications have been reported in the literature, this paper tries to capture the fundamental characteristics of biped walking by limiting our study to this super-abstractive structure.

## 2.  Gait cycle

The complete gait cycle of human walking consists largely of two successive phases: the double-support phase (DSP) and the single-support phase (SSP) with intermediate sub-phases [24, 25]. The DSP arises when both feet contact the ground, resulting in a closed-chain mechanism, and the SSP starts when the rear foot is not supported by the ground and the front foot is flat on the ground. It should be noted that the DSP accounts for about 20% of the time taken for one stride of the gait cycle and the SSP about 80% [16, 25].

Because of the complexity of biped mechanisms, most researchers have simplified the gait cycle to explore their kinematics, biomechanics and control schemes. For details on different walking patterns of biped robot, see [23] and the references therein.

## 3.  Stability

The biped mechanism is unstable during the SSP. One of the challenges in the design and control of biped robots is to maintain their balance while walking in different kinds of environment. The reason for the instability is under-actuation owed to the passive joint of the foot-ground contact. This means that control of the feet is dependent on control of the mechanism above the feet [14, 29]. Common stability theories such as analysis of eigenvalues, gain and phase margins, and Lyapunov stability can be applied to particular modes of biped robot gait but cannot guarantee biped stability for all modes of motion [30].

In general, there are two types of stability criteria that the trajectories of a biped mechanism depend on: static stability and dynamic stability. Static stability restricts the vertical projection of the centre of the mass of the biped to the inside of the support polygon. The support polygon is defined as the area represented by the stance foot during the SSP and the bounded area between the supported feet during the DSP [18]. This type of stability leads to slow gait and biped robots with large feet [31]. Thus, the position of the centre of mass, $\boldsymbol{p}_{com}$, can be calculated as

$$\boldsymbol{p}_{com} = \frac{\sum_{i=1}^{n_l} m_i \boldsymbol{p}_i}{\sum_{i=1}^{n_l} m_i} \qquad (1)$$

where $n_l$ is the number of biped links, $m_i$ is the mass of link (i) and $\boldsymbol{p}_i$ is the position of the centre of mass of link (i). The ground projection of $\boldsymbol{p}_{com}$ can be found easily by determining its components.

Dynamic stability provides more freedom than static stability since the projected centre of mass of the biped may leave the support polygon and thus allow for faster gait [31]. There are four specific techniques for analysing dynamic stability [30, 32-34]:

(1)    Zero-moment point (ZMP), (2) Centroidal angular momentum, (3) Footstep-based criteria (4) Periodicity-based gait.
The first three criteria will be described in details due to their association with humanoid biped locomotion. The periodicity–based gait is not locally stable such that the biped cannot stop, turn etc; therefore, it will not be considered in this paper.

### 3.1 Zero-Moment Point (ZMP)

The notion of the ZMP was proposed by Vukobratovic and colleagues [207,35], who exploited the passive joint of foot-ground contact. It is applied to the biped mechanism in the design of walking patterns and control schemes. The ZMP is the point on the ground at which the net moment vector of the inertial and gravitational forces of the entire body has zero components in the horizontal planes [14, 29]. In brief, if ZMP is located inside the support polygon, then the system is stable and the centre of pressure (COP) of the foot coincides with the ZMP. If the ZMP is outside the support polygon, the system is unstable and the ZMP will be outside the stability margin comprising the fictitious ZMP (FZMP) [14], as shown in Fig. 1. The point P in the mentioned figure represents the location of the zero components of the net moments affecting the foot on the horizontal planes. It is clear that in the stable case one can determine the position of the ZMP by calculating the position of the COP. This is done by using force sensors at the sole plate of the foot. In effect, the theoretical calculation of the ZMP can be performed by the following two formulations:

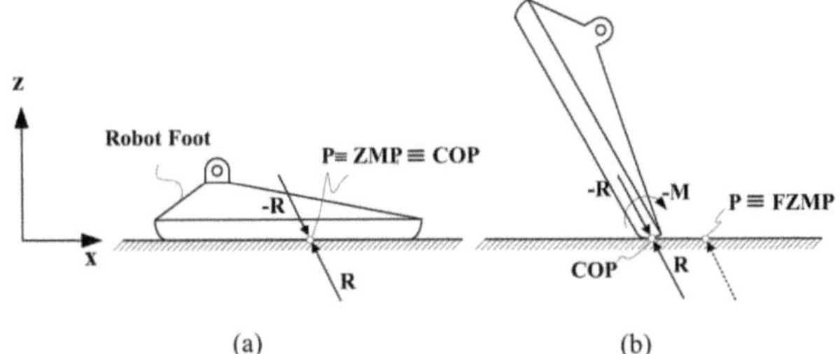


Fig. 1 Relationship between the ZMP, FZMP and COP. (a) Dynamically stable (b) Dynamically unstable [14]

(a) Formulation 1 (more computational)

By knowing the inertial and gravitational forces, one can find the ZMP coordinate as in Eqs. (2) and (3) [36]

$$x_{ZMP} = \frac{\sum_{i=1}^{n} m_i(\ddot{z}_i+g)x_i - \sum_{i=1}^{n} m_i\ddot{x}_i z_i - \sum_{i=1}^{n} I_{iy}\ddot{q}_{iy}}{\sum_{i=1}^{n}(\ddot{z}_i+g)} \tag{2}$$

$$y_{ZMP} = \frac{\sum_{i=1}^{n} m_i(\ddot{z}_i+g)y_i - \sum_{i=1}^{n} m_i\ddot{y}_i z_i + \sum_{i=1}^{n} I_{ix}\ddot{q}_{ix}}{\sum_{i=1}^{n}(\ddot{z}_i+g)} \tag{3}$$

where x and y axes are parallel to the horizontal plane for the biped motion, $z$ axis is pointing upwards, $(I_{ix}, I_{iy})^T$ is the inertial vector of link (i), $\ddot{q}_{ix}$ and $\ddot{q}_{iy}$ are the angular acceleration of link (i) about x and y respectively, $g$ is the gravitational acceleration, $(x_{ZMP}, y_{ZMP}, 0)$ is the coordinate of the ZMP and $(x_i, y_i, z_i)$ is the coordinate of the mass centre of link (i). The above equations can be used to investigate the stability of the biped mechanism during the SSP and DSP.

(b) Formulation 2 (less computational)

Equations (2) and (3) are highly nonlinear; therefore, most researchers use Eqs. (4) and (5) alternately during the SSP, using the static equations of the fixed stance foot [15].

$$x_{ZMP} = -\tau_y/F_z \tag{4}$$
$$y_{ZMP} = \tau_x/F_z \tag{5}$$

where $\tau_y$ and $\tau_x$ are the ankle joint torques about the referred axes and $F_z$ is the normal component of the ground reaction force. Because the ZMP coincides with the COP, the ZMP coordinate can be found during the DSP according to Eq. (6) [38]

$$cop = cop_f \frac{F_{zf}}{F_{zf}+F_{zr}} + cop_r \frac{F_{zr}}{F_{zf}+F_{zr}} \tag{6}$$

where $cop$, $cop_f$ and $cop_r$ represent the centre of pressure for the biped robot during the DSP, the front foot COP and the rear foot COP respectively, and $F_{zf}$ and $F_{zr}$ are the normal components of the ground reaction forces for the front and rear foot respectively.

In general, the ZMP-based biped mechanisms are characterized by the following.

- The stance foot of the biped should remain in full contact all the time [39].
- All the joints of the biped mechanism are actuated and often rigidly controlled to track predetermined trajectories which can simplify the control task during the SSP.
- They do not exploit natural dynamics, so their locomotion usually tends to be less natural than human-like motion (often bent-knee walking) and with high energy consumption.
- Most conventional humanoid biped robots are driven by electric motors via gears and their walking patterns are based on the linear inverted pendulum mode.

Several algorithms have been used to generate a stable reference trajectory for the biped mechanism which satisfies the ZMP constraint. This implies that the solution of Eqs. (2) and (3) to find a relationship between the centre of mass (COM) of the biped and the ZMP trajectory requires a lot of computational effort. Therefore these algorithms can be applied offline [42, 43]. Alternatively, most researchers use simple models to generate the desired biped walking patterns such as the linear inverted pendulum for online implementation. Biped robot stability approaches based on the ZMP still lack efficiency, robustness, easy handling and natural motion [39, 44]. For further reading, refer to references [27, 45-54].

3.2 Centroidal angular momentum

According to the principle of dynamics, the rate of change of linear/angular momentum of a rigid body about its centre of gravity G is equal to the resultant external forces/moments. As a result the linear and angular momentums are zero for zero external disturbances (forces and moments). Exploiting this property, the angular momentum can be used as an index for rotational posture control of biped robots and even for legged systems [55]. Sano and Furusho [56, 57] proposed using the angular momentum of the biped system as a controlled variable. It has been shown that the angular momentum during the SSP about the ankle joint of the stance foot is a function of the gravity effect and the ankle torque. Therefore, the ankle torque can be considered as an angular momentum control input. In their method, the torque control rather than the position control have been used for the ankle joints during the SSP and one of the hip joints during the DSP with specific selection of the desired angular momentum function. Goswami and Kallem [55] suggested an important criterion called the zero rate of change of angular momentum (ZRAM) point. They observe that the rate of change of centroidal angular momentum is a useful index/criterion for the analysis and control of posture control of biped robots in different interactive environments. Figure 2 shows the detail of their method.

They proposed three control strategies for balance control, manipulating the value of the rate of change of the angular momentum as follows.

$$OP \times R + OG \times m_{COM} \times g = \dot{H}_G + OG \times m_{COM}a \qquad (7)$$

with notations shown in Fig. 2.
These strategies can be applied through the following:
1. Enlarging the support polygon.
2. Moving $G$ relative to point P; this coincides with strategy of COG position-based compensation discussed in Table III.
3. Changing the direction of ground reaction forces by changing the centroidal (spin) acceleration of the biped mechanism; this coincides with the strategy of COG acceleration-based compensation discussed in Table III.

*Remark 2.* Although some researchers have not directly considered the angular momentum as a balance criterion, the latter two strategies have been used successfully to compensate for modelling errors and external disturbances, as we will see later. In addition, sudden forward movements of the body and arms make use of the angular momentum to maintain balance [32]. Therefore, all the methods described in Section 5.2 try to regulate the centroidal angular momentum.
Popov and colleagues [54, 58-61] proposed the centroidal moment pivot CMP as a stability criterion for measuring the unbalance of the biped mechanism. It is exactly the same criterion proposed by [55]. It is the point where a line parallel to the ground reaction force, passing through the centre of mass, intersects with the external contact surface [54], as shown in Fig. 2.

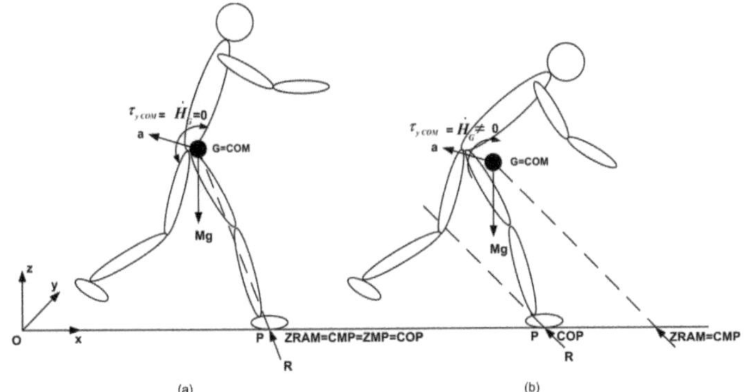


Fig. 2 The relationship between ZMP, CMP and ZRAM points. (a) Dynamically stable biped (b) Dynamically unstable biped [55,54]

It can be written in terms of the centre of mass and ground reaction forces as in (8) and (9) and in terms of the ZMP location, the vertical ground reaction force, and the moment about the centre of mass as in (10) and (11).

$$x_{CMP} = x_{COM} - \frac{F_x}{F_z + m_{COM}g} z_{COM} \qquad (8)$$

$$y_{CMP} = y_{COM} - \frac{F_y}{F_z + m_{COM}g} z_{COM} \qquad (9)$$

$$x_{CMP} = x_{ZMP} + \frac{\tau_{yCOM}}{F_z + m_{COM}g} \qquad (10)$$

$$y_{CMP} = y_{ZMP} - \frac{\tau_{xCOM}}{F_z + m_{COM}g} \qquad (11)$$

where $(x_{COM}, y_{COM}, z_{COM})$ is the position of COM, and other notations are shown in Fig. 2
It is clear from (10) and (11) that if the centroidal moments of the biped are zero, the CMP will coincide with the ZMP and the biped is stable (balanced). This now means that the centroidal moments should be zero for a balanced biped; corrective moments for minimizing the non-zero spin angular momentum should be applied in the case of unbalanced bipeds. As a result, the ZMP may not coincide with the CMP in some cases. One of the important problems inherent in the notion of ZMP which motivates researchers to look for a general powerful stability index represented by angular momentum is that of the point-foot biped robot. In that case the ZMP is constrained at the contact point and cannot be repositioned. The only way to stabilize such a system is to produce a non-zero moment about the COM of the biped. Kajita and colleagues [62, 63] used the ZMP to generate reasonable walking patterns and integrated the notion of specific angular momentum as a high-level control to maintain the balance of the biped mechanism.

*Remark 3.* Most common humanoid robots depend on the ZMP in generating stable walking patterns. Motion planning is applied with the ZMP and the balance strategy can be performed by regulating angular momentum and the ZMP stabilizer. In other words, as proved in [63], the ZMP can be maintained inside the support polygon by selection of suitable values of the total linear and angular momentum and the position of the COM.

*Remark 4.* The foot rotation indicator (FRI) [29] and FZMP are equivalent terms which can be used to measure the unbalance of the biped mechanism without explaining their relationships with the linear/angular momentums, except in the work of [63]. The ZRAM point and CMP are exactly equivalent in terms of regulating the angular momentum in relation to the COG. Even now there is no general criterion that can describe all stability cases of biped locomotion with different gaits [55].

### 3.3 Footstep-based criteria

Pratt and Tedrake [32] proposed five subtasks that the biped should perform to generate balanced gait: (1) maintaining the body (trunk) orientation, (2) maintaining virtual leg length resulting from simplified linear inverted pendulum plus flywheel body, (3) swinging the swing leg, (4) transition of different phases, and (5) regulating the velocity of the centre of mass. They called them velocity-based stability margins. Point 5 plays an important role in maintaining balance. It has been shown that the first three points are fully controllable and can be implemented by means of traditional high gain joint position control techniques, provided the requirements of joint torques are satisfied to avoid slipping. The fourth point can be dealt with by low impedance force control [64], or a shift transition function that guarantees smooth transition of the ground reaction forces from the rear foot to the front one [65]. The most challenging problem is point 5 such that the COM velocity, beyond a certain point, is not controllable once the projection of the COM moves away from the support foot. The only possible strategy in this case is to take a step forward to maintain balance. Therefore, capture points have been proposed [33] to make a step for balance maintenance. A capture point is a point on the ground that can be stepped on by the biped robot in order to stop. Pratt and colleagues [33, 66] extended the linear inverted pendulum to include a flywheel body at its COM to explicitly model angular momentum about the COM. A closed form solution of the capture region was devised. Their model is approximate, however, and its application to complex humanoid robots is questionable! Wight *et. al.* [34] proposed a dynamic measure of balance to estimate the footstep to restore balance. The ZMP criteria can be augmented with this dynamic measure to broaden the stability conditions of the biped as it walks.

Comparison between the above mentioned stability criteria are described in Table I.

Table I Comparison of different stability criteria [32, 33]

| Stability Criterion | Advantages | Disadvantages |
| --- | --- | --- |
| ZMP | It is a sufficient and conservative stability criterion for biped robots which are fully actuated and position controlled to track predetermined trajectories. | 1. The stance foot of the biped should remain in full contact at all times. <br> 2. Unnatural motion with high energy consumption. <br> 3. There are many situations in which the ZMP is violated, e.g. during toe-off human walking, biped with point feet, passive dynamic walkers. On the other hand, the ZMP could be satisfied while the biped robot is collapsed, e.g., setting zero torques of the biped mechanism and maintaining a flat foot while hitting the ground with its trunk. |
| Centroidal angular momentum | It is a good tool for recovering from a push or other disturbance, as there is a coupling between the angular momentum rate change and linear angular momentum and hence the COM speed. | Depends on minimizing spin angular momentum and can be violated through walking with a violently thrashing upper body, and falling over while the COM has zero spin angular momentum. |
| Footstep-based criteria | One of the successful strategies for recovering from push. | 1. Computation of capture points may be difficult and obtaining a closed form solution might not be possible for a general humanoid robot. <br> 2. Depends on an approximate model. |

## 4. ZMP-based control

Human locomotion on level terrain with constant speed walking is controlled by a local pattern generator on the spinal cord without the use of brain commands. The control system of human walking in a cluttered environment is more complex because this may involve multiple local controllers and commands from the cerebellum [3]. Maintaining human balance while walking, standing, etc. is a complex process associated with different levels of the central nervous system (CNS) [67]. In general, balance depends on the task characteristics and environment context as detailed in [68]. The control system which is responsible for human walking can be described by hierarchical controls ranging from the highest level of action planning (motion planning) to the low-level control (reflex/local control) [3].

In robotics, multiple control loops have been used efficiently for generating feasible walking for biped mechanisms. As mentioned earlier, most ZMP-based humanoid robots need to use multi-level control to adapt to changes in the environment and internal perturbations. In contrast, the periodicity-based biped does not afford robust walking and cannot stop or stand etc; therefore, using a balance control is challenging. Figure 3 shows the classification of control levels of general biped mechanisms.

*Remark 5.* Offline walking patterns do not need multi-level controllers, because they have been prepared in advance to satisfy the dynamic/kinematic constraints [65, 69, 70, 102]. It needs only low-level control to track the desired angular joint trajectories.

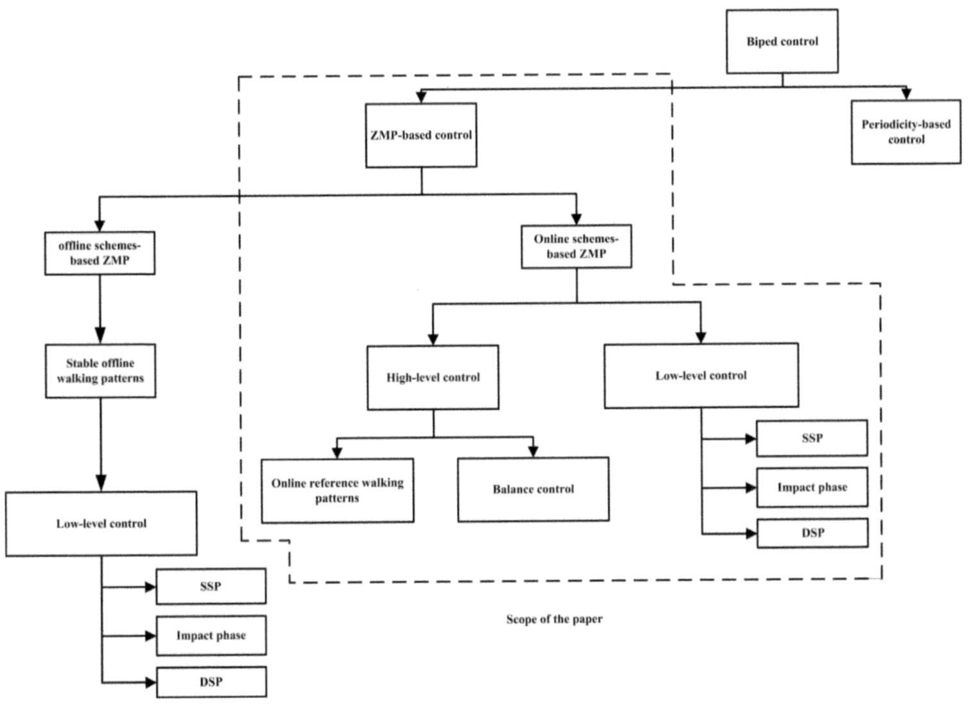

Fig. 3 General classification of multi-level control of biped mechanism according to its stability criterion.

## 5. High-level control

Our paper suggests that the control systems of the biped mechanism consist of two levels: (1) high-level control which is responsible for motion planning and balance control, and (2) low-level control which can deal with the tracking of the desired angular joint trajectories.

### 5.1. Online reference walking patterns

Generating walking patterns for biped robots demands finding a relationship between COG and ZMP trajectories. Depending on the reference trajectories of the COG and the feet, all the motions of the joints can be deduced. [15]. Numerous approaches have been used to generate the motion of biped robots but only three methods have been used successfully: optimization-based gait, COG-based gait and interpolation-based gait. The optimization tool can deal successfully with minimum energy or input control effort, optimal design and miscellaneous dynamic and kinematic constraints [65, 69, 70, 102]. ZMP criteria can be seen as a constraint or an objective function depending on the designer's purpose. Because of the computational complexity of the optimization tool it can often be used offline. The COG-based gait adopts the inverted pendulum model as a simplified model for biped robots, assuming all masses are concentrated in the COG and there is a pushing force at the ankle joint of the support foot without applied torque [71]. This model has been suggested for online realization of walking patterns for the biped robot. To compensate for the error produced by this approximation and other unknown disturbances, modification of the walking patterns is necessary via a balance control strategy which will be described later. The so-called interpolation-based gait requires suitable polynomial or spline functions for the COG and the foot trajectory such that it can track the desired ZMP or just satisfy it [21, 26, 27, 72-76] with offline/online compensated ZMP. The method of Takanishi et. al. [76] can be categorized within interpolation-based gait because the authors used fast Fourier series to find a suitable relationship between the trunk motion and ZMP trajectory. Because our paper focuses on online stable biped walking, we will examine the COG-based gait in detail. Real-time motion planning is difficult to perform for the following reasons: (1) the proposed algorithm for the walking generator should be solved in one iteration otherwise offline algorithms should be used, and (2) it is difficult to connect the predefined walking pattern smoothly online because of the discontinuous COG velocity in the transition instances [77]. The online walking pattern generators can be satisfied by tuning input parameters simultaneously as desired. These parameters could be sway amplitude, step time, stride, etc., which are prepared before the step starts and updated at the start of each step [78]. For more details about walking pattern generation, refer to [23, 79-89].

Because of the complex nonlinear dynamics of the biped robot, it is difficult to obtain a closed form solution. Therefore, many researchers have focused their attention on simplifying the complex dynamics of biped robots in order to use simple algorithms for generating walking patterns and control. As mentioned earlier, the inverted pendulum model is used as an approximate model for the biped mechanism because of the similarity between the two mechanical systems in terms of both the static and the periodic stability. This approximation can be connected with the periodic stability of the inverted pendulum, however. Kajita and Toni [71, 90] proposed the linear inverted pendulum mode (LIPM), exploiting previous related work [13, 91-97]. The LIPM includes finding a simple mathematical relationship between the trajectory of the centre of gravity (COG) of the biped robot and the ZMP to ensure stable biped motion. It is assumed that all masses of the biped model are concentrated in its COG and there is a reaction force between the ground and the foot without ankle torque [43, 98, 99]. Thus, the relationship between the COG and ZMP during the SSP can be computed as [15, 100-102]

$$x_{ZMP} = x_{COG} - \frac{H}{g}\ddot{x}_{COG} \qquad\qquad (12)$$

$$y_{ZMP} = y_{COG} - \frac{H}{g}\ddot{y}_{COG} \qquad\qquad (13)$$

where $(x_{COG}, y_{COG})$ is the position of the COG, and $H$ is the height of the COG which is assumed to be fixed.

*Remark 6.* Modelling of the biped robot during the DSP can be performed using different strategies: (1) the linear inverted pendulum model [101], (2) the linear pendulum model [100], and (3) suitable transition function connecting the end of the first SSP to the beginning of the second SSP; it could be linear function or the same equation of the SSP but with opposite sign as performed in [15]. Consequently, the global behavior of the bipeds could depend on pendulum model strategy during the complete gait cycle. Equivalence of these methods has been proved in [102].

*Remark 7.* Certainly there are miscellaneous methods used for generating walking patterns of the biped as detailed in our previous paper [23]; however, we focus our study on COG-based walking pattern to capture the most important issues of balance control associated with biped locomotion.

*Remark 8.* In order to decouple sagittal and frontal motion, hip height needs to be constant and avoid singularity at the knee joint. But then the robot walks with bent knees. For further reading on stretch-legged walking, refer to [28, 103-105].

The LIPM concept could be accepted for quadrupeds because of their slender legs. One-third of the mass of the human being is contained in the legs, however. Therefore, the use of the LIPM in bipeds is questionable [16]. Consequently, Park and Kim [43] suggested using the gravity-compensated inverted pendulum mode (GCIPM) with two proposed masses to reduce the errors accompanied by the LIPM. One mass denotes the dynamics of the swing leg and the other represents the stance foot and the trunk. Other modifications to the LIPM have been proposed to avoid the deviation of the desired trajectories of ZMP. These modified approaches include the two masses inverted pendulum mode (TMIPM) [106], the multiple masses inverted pendulum mode (MMIPM) [107], and the virtual height inverted pendulum mode (VHIPM) [108]. For a comprehensive comparison of the mentioned methods, refer to Ref. [108]. Table II shows in details most of the methods used for online trajectory generation of the biped gait.

Table II Detailed description of the methods and comments on online reference walking patterns

| Method | Description | Comments |
|---|---|---|
| Indirect ZMP control | <ul><li>It uses the inverted pendulum to approximate the dynamics of the multi-body biped robot (see Fig.4 (a)).</li><li>The ZMP appears as the input of the system's plant.</li><li>The controller attempts to track the desired COG position /velocity.</li><li>Example of this controller can be seen in Fig. 5.</li></ul> | <ul><li>The problem of the non-minimum phase property is inherent in the one-mass inverted pendulum (or the cart-table model) [111]. It can be solved either using two-mass inverted pendulum [111] or pole-zero cancellation [112].</li><li>It could be possible to use any linear control technique (PID controller, $H_\infty$ control etc.) for control of inverted pendulum model (or the cart-table model) to track the desired ZMP or COG states [109]. However, preview control and its extensions [113, 115-117] and linear predictive control [118-121] have been used extensively.</li><li>The ordinal control system cannot make good tracking for ZMP trajectory as shown in [15]. Moreover, the disadvantages of the preview control mentioned above are: (1) it has difficulties in specifying the desired ZMP precisely at the heel or toe of the foot to realize human-like walking [114], (2) it does not consider the constraints of the ZMP, and (3) it is sensitive to disturbance and noise. Therefore, linear predictive control has been proposed to deal with strong perturbations [118-121].</li></ul> |
| Direct ZMP control | <ul><li>It uses the cart-table model to to approximate the dynamics of the multi-body biped robot (see Fig.4 (b)).</li><li>The ZMP is the output of the system.</li><li>Example of this controller can be seen in Fig. 6.</li></ul> | |

| | | |
|---|---|---|
| | | • Arbulu *et. al.*[122] have suggested, without proof, that using the cart-table model can guarantee high walking speed with continuous dynamic response.<br>• Large computational effort is often accompanied by most ZMP/COG tracking control methods [15]. |
| Analytical method | • Apart from the control theory, the researchers have solved the Eqs. (12) and (13) analytically using two successful techniques: (1) the two-point boundary value problem (BVP), and (2) the convergent and divergent components of the motion[114].<br>• The two-point boundary value problem can be solved by expressing the desired ZMP trajectory as a function (polynomial, Fourier series functions, etc.) and then determining the COG position and velocity to achieve the desired ZMP trajectory (please see Refs. [77, 78, 124-129]).<br>• Instead of the BVP, Takenake *et. al.* [114] have proposed to decompose motions of the inverted pendulum into convergent and divergent components. Modification of the ZMP can be decreased by only controlling the divergent components. | • In general, the analytical method can be easy and less computational than other control methods mentioned above.<br>• For detailed discussion of disadvantages of the BVP, see Ref. [114]. |

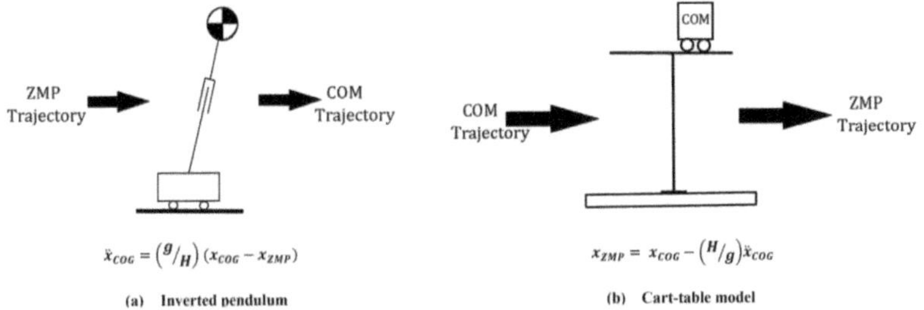



$$\ddot{x}_{COG} = \left(g/H\right)\left(x_{COG} - x_{ZMP}\right)$$

$$x_{ZMP} = x_{COG} - \left(H/g\right)\ddot{x}_{COG}$$

Fig. 4 The inverted pendulum model vs. the cart-table system[122]

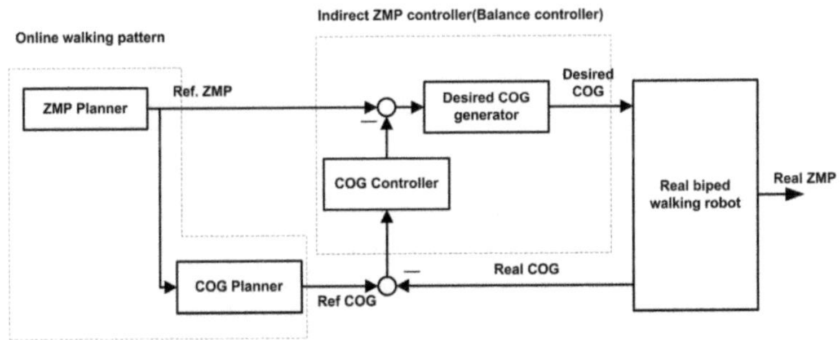



Fig. 5 Indirect control of ZMP-based biped robot for tracking of desired COG position [110].

*Remark 9.* Assuming the ZMP trajectory during the SSP is fixed at some point of contact could have undesirable effects on biped motion. It can limit the performance of the humanoid robot; it cannot vary step length or walking period, etc. in the same way a human does. Therefore, linear transition of the ZMP is preferred for achieving some level of control of COM motions [130, 131]. Vanderborght *et. al.*[15] showed that the acceleration of variable ZMP is smaller than fixed ZMP; however, the ZMP comes closer to the boundary of the support area. In effect, more work is needed to determine the optimal trajectory of ZMP with satisfied constraints.

## 5.2 Balance control

All legged creatures have amazing balance strategies to avoid falling suddenly and to maintain their overall rotational stability, dynamic stability and postural stability [55].Humans can use a variety of complex strategies in order to maintain balance and avoid potential disturbances. The balance system proactively monitors the external environment and predicts the effect of forces generated by voluntary movements on the body, making the adjustments necessary to maintain posture and equilibrium. It is only when these adjustments fail or unexpected destabilization occurs that reactive balance response emerges [68]. Thus, two familiar mechanisms are used to achieve balance conditions, the reactive mechanism (feedback control) and the proactive mechanism (feedforward control) [67]. Table III describes in details the four known strategies for balance control of biped locomotion.

*Remark 10.* Reactive response occurs after unexpected disturbances are applied to the human [67]. In such cases, the human may use three essential strategies to maintain balance according to the amount of the disturbed forces applied to the body. For small disturbances, the impact is absorbed by the ankle joint; moving the COM of the human body forward or backward as a flexible inverted pendulum about the ankle joint [67, 130]. This has motivated the use of an inverted pendulum with virtual spring and damper elements to excite the balance control using ankle strategy. When the impact is larger, the hip and knee joints are used such that the body is stiffened as a two-segment inverted pendulum. Whereas for severe and fast perturbations or for maintaining vertical trunk orientation, stepping the foot could be the best solution (see Fig. 7). For further details about these strategies refer to [131-134].

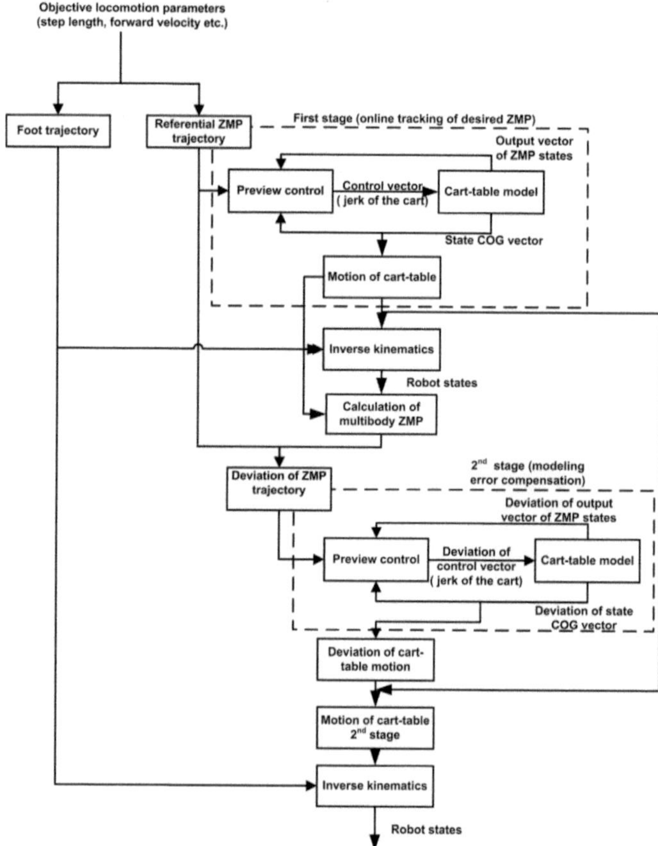


Fig. 6 The two stages of the preview controller for Lucy biped [15]. Stage one can track the desired predicted ZMP trajectory and the second stage compensates for the ZMP errors produced by the differences between the simplified model and the actual multi-body model.

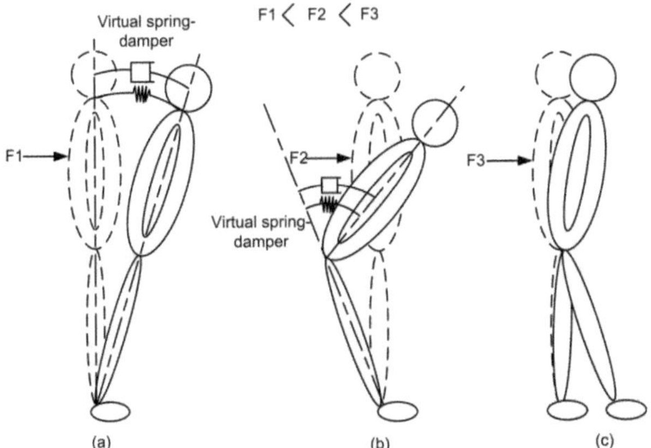


Fig. 7 The familiar balance strategies. (a) Ankle strategy with physical modelling (b) Hip strategy with physical modelling (c) Footstep strategy [67, 135-137]

Table III Balance control and stabilization strategies of biped robots

| Method | Description | Comments |
|---|---|---|
| Ankle-based strategy (reactive mechanism) | • The idea is to make the ankle joint of the stance leg to behave as passive joint. Then the deviations of the desired ZMP could be small as possible[138]. <br>• In general, the compensating ankle torque can be modelled as a rotational spring and damper at foot-ground contact (see Fig. 7 and Refs. [78, 138]), or it can be computed from the force sensors located at biped feet (see Ref. [139]). | • A compliant ankle joint makes the foot landing easy and gives sufficient contact to the ground [208]. However, adding flexibility to the ankle joint could complicate the required control system because of the possible resonant oscillations (successful control architecture for flexible ankle joint based on ZMP criterion can be found in [141]). <br>• Due to the inherent instability of the biped and other external disturbances, the ankle compensation control could not be enough; therefore, ankle-knee compensation control [137], ankle-hip compensation control [135,136, 140] strategies have been proposed. |
| Hip-based strategy (reactive mechanism) | • This strategy is more robust than the ankle strategy (see Fig. 7). <br>• One possible solution to compensate for the ZMP deviations is modifying the orientation of the large-mass trunk while the lower limbs have a predefine trajectory (see the work of Takanishi and his colleagues [35, 76, 107]). <br>• To avoid the high energy consumption of the large trunk, COG position and acceleration compensation methods have been proposed (see Figs. 8 and 9). The idea is to modify the position or the acceleration or both of them to compensate for ZMP errors. | • The heavy trunk strategy may lead to high energy consumption with possible difficulties of control of the upper limb [27]. <br>• It has been shown that the COG position compensation method can be applied within low frequency value whereas COG acceleration-based compensation method can work with a high frequency range [142]. <br>• It should be noted that the mentioned COG-based compensation methods do not incorporate control of the ground reaction forces to avoid slipping (see Fig. 10 which incorporates the effect of ground reaction forces). |
| Whole-body strategy (reactive mechanism) | • This strategy is based on modifying the whole body of the biped (legs, torso, arms and head). It is more robust than the hip-based compensation method. <br>• There are different methods used in the literature for the whole body modification as follows: resolved angular momentum [62] (see Fig. 11), the angular momentum inducing inverted pendulum mode (AMPM) [101, 130], Jacobi-based technique [145-147], filter dynamics [148-150], passivity theory [40, 152-155]. | • The position-based control methods (resolved angular momentum, AMPM, Jacobi technique and filter dynamics) cannot quickly adapt to sudden environment changes, especially unknown external forces [49]. In addition, a position-based compliance control requires force measurements at every expected contact point, which increases the computational load and produces unavoidable time delays [41] (please see Refs. [40,41] for more details). <br>• The ground reaction forces are the key elements through which new control strategies are proposed to provide the level of compliance, adaptation and |

| | | dynamic stability required for walking in different contexts [151]. Therefore, modification of the biped gait for stabilization purposes combined with control of ground reaction forces is important (see e.g., passivity theory [40], hybrid position-force control [151]). |
|---|---|---|
| Footstep strategy (reactive mechanism) | • If the control of COG velocity is impossible, the only possible way to recover balance is take a step forward.<br>• For details on the characteristics of this strategy, see Section 3.3. | • A large external force cannot be easily suppressed by any state feedback, especially if applied statically for a long time, because of the horizontal motion of the trunk. On the other hand, it is difficult for the swing leg to execute a large motion just before landing. Furthermore, if an extra step is used to compensate for a small external force without using any feedback suppression, the robot moves throughout the step and consumes a lot of energy.Therefore,it is logical to distribute the action each way to adequately compensate for the disturbance, according to its frequency or the situation of the support phase [128]. |
| Predictive control (proactive mechanism) | • The locomotion of the human being on cluttered environments without prescribed reference trajectory in advance motivates the researchers to use the predictive control for biped locomotion [156].<br>• The model predictive control is a branch of optimal control that needs objective function for the walking patterns, e.g. control effort, to satisfy given constraints, e.g. hip orientation, knee locking, foot trajectory, etc., as shown in [118, 119, 157-161]. | • Two problems have been noticed with the application of Trajectory Free –Model Predictive Control (TF-MPC) in real time, which are the computational complexity and the size of the horizon; see Refs. [162, 119-121] for solutions of computational complexity.<br>• The MPC is sensitive to any uncertainty in biped model parameters or external disturbances; therefore, the disturbance observer can be incorporated with it as made in [163, 164]. However, the computational complexity in Refs. [163, 164] was not considered and therefore its real-time application is questionable. |

*Remark 11.* As we see from Table III, the robotics community has discussed extensively the problem of the balance control of the biped robot, but in most cases the balance control strategies (ankle, hip strategies etc.) mentioned above are dependently studied without integration of them. However, some of these balance strategies have been integrated as made in Refs. [128, 129, 135-137, 140].

*Remark 12.* Optimal predictive control combines two balance strategies: a reactive mechanism and a proactive mechanism. In addition, it can directly determine the desired input controls without reference trajectories, incorporating low-level control implicitly. Therefore, it could be a promising strategy for control and balance of biped robots taking the computational complexity into consideration.

*Remark 13.* Intelligent control techniques have attracted the attention of many researchers wishing to generate and stabilize biped walking patterns. Katic and Vukobratovic [165] performed a useful survey about these useful techniques for controlling biped humanoid robots.

The results presented in Section 5 can be summarized as follows:

1. If the biped robot moves without ZMP feedback, any perturbation could lead to imbalance [166].

2. The essential idea is to keep the ZMP inside support polygon. In order to control it there are several possibilities: (1) change the ankle torque directly, (2) change the hip motion or flap with arms around (like we do when balancing on a cord), (3) combination of points (1) and (2) which is called whole body control. (4) footstep strategy (If the ZMP tends to go outside support polygon, the above strategies could be insufficient anymore; so, we need to make a step to recover).

3. It has been observed that it is possible for a rotation of the whole body of biped mechanism while walking with zero spin angular momentum, or having non-zero balancing angular momentum while walking stably. With these observations, the assumption of [53, 54, 58-61] to balance biped mechanism by tracking zero angular momentum is questionable; instable biped locomotion could be come from that assumption. On the other hand, tracking specific angular momentum as made in [62, 63] could be acceptable and accompanied by difficulties in selection of this non-zero value. Therefore, varying step length of biped walking (footstep) or walking speed could be safe strategy (in some circumstances) for stabilization and balance purposes [167].

4. The footstep technique can make instability problems if the biped must move on certain locations (e.g. stepping on stones) [166].

5. In some situations, controlling the ZMP trajectory is not sufficient to maintain dynamic stability of the biped mechanism such as moving on smooth surfaces. In that case, it is important to control the ground reaction wrench (forces and moments) as well as ZMP trajectory together to ensure stable and non-sliding motion [143, 144, 166].

6. Optimal predictive control considering computational complexity can be used as global approach for stabilization of biped robots.

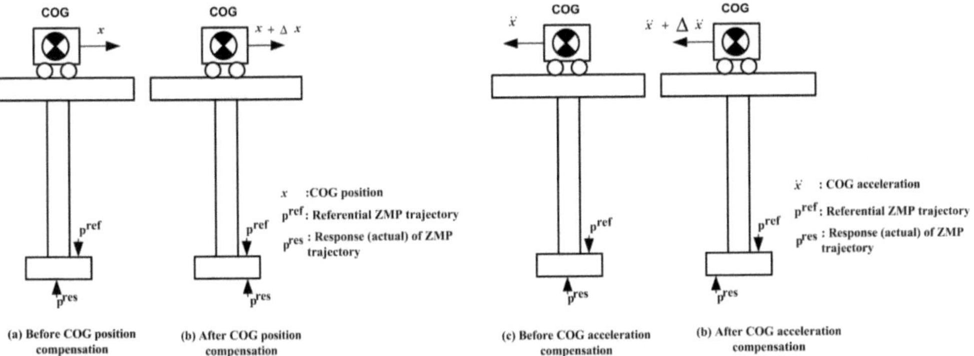



Fig. 8 COG-based compensation. (a) and (b) Before and after COG position compensation respectively (c) and (d) Before and after COG acceleration compensation respectively [142].

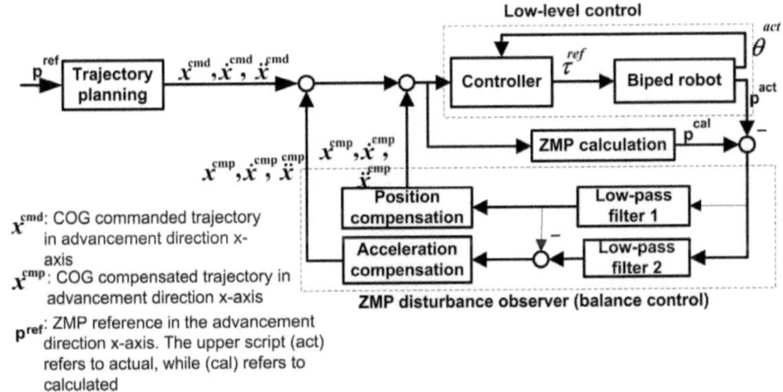



Fig. 9 ZMP disturbance observer [142]

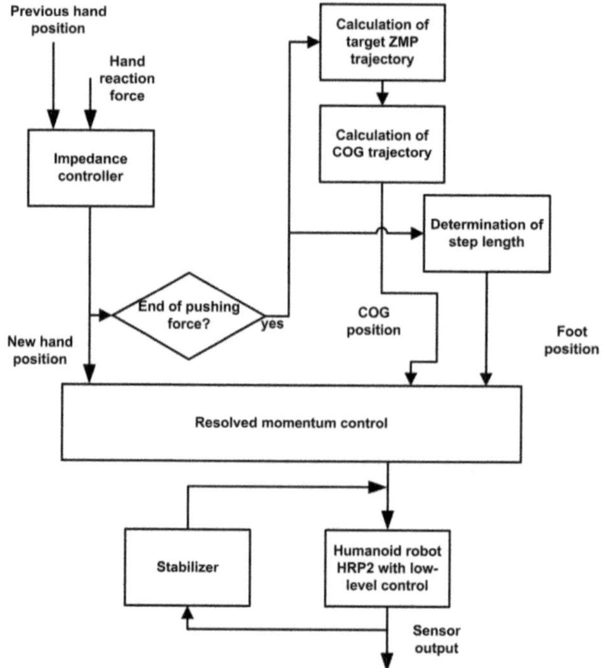



Fig. 10 Multi-level control of ASIMO humanoid robot[143,144]



Fig. 11 Multi-level control of HRP2 humanoid robot[77]

## 6. Low-level control

Tracking the desired angular joint trajectories of the biped mechanism is the task of the low-level control. As mentioned earlier, the notion of the ZMP criterion assumes a fully actuated biped robot during the SSP to track the desired trajectory of the ZMP or satisfy it and over-actuated biped during the DSP. Consequently, the stance foot of the biped robot is assumed to be fixed during the SSP and the conventional control techniques used for controlling manipulators can be used successfully. In effect, most of the methods investigated in the excellent work by [168-170] can be used sufficiently to control ZMP-based gait. The designer should, however, consider the high degrees of freedom inherent in the biped humanoid robot. Therefore, most humanoid robots have simple local PID controllers for tracking purposes [23]. Below we investigate the control techniques used in biped robots according to their walking phases. Table IV shows with details the control systems used during the walking phases of biped robots.

Table IV Low-level control during different walking phases of the biped locomotion

| Method | Description | Comments |
|---|---|---|
| The SSP | • During the SSP, the biped robot behaves as an open-chain mechanism; therefore, the Lagrangian (or Newtonian Euler [171,206]) dynamic equation can be written as<br><br>$$M\ddot{q} + C\dot{q} + g = A\tau \qquad (14)$$<br><br>where $M \in \mathbb{R}^{n_q \times n_q}$ is the mass robot matrix, $q, \dot{q}$ and $\ddot{q} \in \mathbb{R}^{n_q}$ are the absolute angular displacement, velocity and acceleration of the robot links, $C \in \mathbb{R}^{n_q \times n_q}$ represents the Coriolis and centripetal robot matrix, $g \in \mathbb{R}^{n_q}$ is the gravity vector, $A \in \mathbb{R}^{n_q \times n_\tau}$ is a mapping matrix derived by the principle of the virtual work [214, 215], $\tau \in \mathbb{R}^{n_\tau}$ is the actuating torque vector and $n_\tau$ represents the number of actuators .<br><br>• Most ZMP-based biped robots are fully actuated during the SSP, therefore, all classical control methods used for robotic manipulators can be used successfully (see Ref. [168]).<br><br>• The control methods used in the literature are: linearization [16, 90], PID family control [90, 19], computed torque control [19], sliding mode control and its extensions [19, 173-175, 176-178], learning control [179-181]. | • The disadvantage of the linearization-based control is that the solution is valid only in small regions around the operating points (nominal trajectory) [216].<br><br>• During motion of any walking robot, a number of sudden geometric constraints are imposed, e.g. stepping on the ground, knee locking, etc. These inherent constraints give rise to impulse-like disturbances that make control by a standard PID controller extremely difficult [19] (if the uncertainty level is very high (more than 80%) it may not be possible to maintain stable gait with the usual PID).<br><br>• Most researchers avoid using adaptive control in complex biped mechanisms because the classical adaptive control could depend on a regressor matrix which is applicable to 6-DOFs or fewer. The virtual decomposition control (VDC) proposed by Wen-Hong Zhu [20] is a promising solution to this problem; it virtually decomposes every link of the complex system and applies a similar classical adaptive control to each link subsystem[128]. |
| The impact phase (initial contact) | • During the impact phase, the robot configuration remains unchanged with abrupt change in the joint velocity. The dynamic equation can be written as follows:<br><br>$$M(q)(\dot{q}^+ - \dot{q}^-) = J^T \int_{t_o}^{t_o+\Delta t} \lambda \, dt \qquad (15)$$<br>$$J^T(\dot{q}^+ - \dot{q}^-) = 0 \qquad (16)$$<br><br>where $\dot{q}^+$ and $\dot{q}^-$ refer to the angular velocity vector of the biped mechanism after and before impact respectively, $J \in \mathbb{R}^{n_\lambda \times n_q}$ represents the Jacobian matrix resulting from the constrained motion of the biped robot, $\lambda \in \mathbb{R}^{n_\lambda}$ is the constrained force/moment vector at the foot-ground interaction with dimension of $n_\lambda$, and is the time.<br><br>• The human has amazing compliance control to attenuate the impact effects; her/his muscles are relaxed to absorb the impact/contact just before her/his landing foot contacts the ground while her/his muscles are hardened to maintain balance after landing [183].<br><br>• The mechanisms are usually designed so that the impact forces are reduced by minimizing the impact velocity and the mass of the impacting bodies, and by suitably designing the mechanism so that a minimum stiffness is located in a neighbourhood of the point of impact, but these designs seem to be more useful for reducing the impact effects than precise control [187].<br><br>• Modelling of feet-ground interaction can be divided into two main groups: (1) rigid approach (constraint-based approach/non-smooth dynamics formulation [188]), and (2) complaint contact (spring-damper) model (penalty-based approach/regularized approach [148, 189, 188]). | • If the impact dynamics is not properly modelled and controlled, the impact forces could result in poor system performance and instability.<br><br>• The difficulties of controlling systems subject to impact include: (1) the dynamics is different during system transition from a non-contact state to a contact state, (2) measuring the impact force, which depends on the geometry of the robot, (3) geometry of the environment and (4) type of contact [186].<br><br>• One possible solution to avoid impact is generation of reference trajectory such that the swing foot has zero impact velocity at the impact instance, avoiding impact phenomena [195-197]. However, this strategy and the rigid approach may need high power actuators whereas the compliant contact model can alleviate the energy consumption resulting from impact. Therefore it is used for simulating most designed humanoid biped robots such as Saika-3 [192], Johnnie [193], and OpenHRP [194].<br><br>• The main disadvantages associated with the complaint contact model are parameter tuning and multiple unilateral contacts [189]; efficient and precise simulation of collisions and contacts still remains an open research issue [148]. Other problem is the rubber foot sole of many bipeds that makes the robot turn slightly to the front so that the foot touches earlier than expected.<br><br>• Impedance control can be powerful strategy to control biped robot with compliance (see Refs. [198, 172, 184, 185]). |

| | | |
|---|---|---|
| | • In the rigid approach, the colliding bodies are hard enough to resist deformation [190]. The equation of system motion is complemented by constraint equations associated with the foot position [191]; the impact event (if it exists) should be modelled as a discontinuous event complicating the control solutions (see [188,189] for investigating disadvantages of this method).<br>• The complaint contact model (second approach) uses spring-damper elements between the feet and the ground for modelling contact [191]. | |
| The DSP | • During the DSP, the biped constitutes a closed chain mechanism with redundant coordinates. Therefore, the following Lagrange formulation of the 1$^{nd}$ kind, which can deals with constraints, is needed for dynamic modeling of the constrained biped.<br>$$M\ddot{q} + C\dot{q} + g = A\tau + J^T\lambda \qquad (17)$$<br>$$\varphi(q) = 0 \qquad (18)$$<br>where $\varphi(.)$ represents the constraint equation, with the same notations mentioned earlier.<br>• Eqs. (17) and (18) can be solved using two well-known techniques [203,204]: (1) the redundant coordinates-based techniques which are used mainly in commercial software such as MSC ADAMS, and (2) the minimum coordinates-based techniques which could be, to some extent, suitable for control strategies and real-time applications. | • If the robotic mechanism contacts with environment, it will lose some DoFs. Consequently, the generalized coordinates of the target robot could be larger than its DOFs due to its constrained motion.<br>• Many researchers have preferred the minimum coordinates-based technique due to its simplicity and ease of derivation at the expense of difficulties of numerical methods encountered in the solution [203]. Consequently, this motivates the researchers to investigate the second technique which includes eliminating the constraint equations (Lagrange multipliers) from Eq. (18) to result in constraint-free differential equations [203]. This can be implemented using one of orthogonalization methods which are [204]: coordinate partitioning method, zero-eigenvalue method, singular value decomposition (SVD), QR decomposition, Udwadia-Kabala formulation, PUTD method, and Schur decomposition.<br>• There are several approaches proposed for control of robotic manipulators with constrained motion; for details, refer to [205].<br>• To ensure continuous dynamic response, Some researchers proposed a linear shift function for the distribution of the ground reaction forces such that there is a gradual shift of the body from the rear foot to the front foot [37,66,200,201,206]. |

The results presented in Section 6 can be summarized as follows

1. Although most researchers have used local controller (decentralized controller) such as PID family for controlling complex dynamic systems, the problem is that the independent joint control considers the coupling effects of the other bodies or links as disturbances, which is unrealistic [20]. Dallali *et. al.* [202] have made a comparison between multivariable and decentralized control strategies for 10-DOF humanoid robot. It was shown that the multivariable linear quadratic regulator (LQR), which takes the entire system dynamics into account, has better robustness and energy efficiency.
2. There are two interesting control techniques dealing with uncertain plant: robust control and adaptive control. The robust control law tries to stabilize the target system within assumed bounds; therefore, it may lead to saturation problem. Whereas, adaptive control tries to estimate the unknown parameters to deal with uncertainty problem. Applying these techniques to complex robotic system such as biped robot is difficult due to inherently high DOFs. So, recursive techniques should be used to resolve this dilemma. A promising control system is the VDC based on adaptive technique which deals with every subsystem alone; it can be used as local controllers considering the full dynamics of the complex robotic system.
3. Most humanoid robots prefer using compliant model of foot-ground interaction to solve the problem of over-actuation during the DSP and to exploit it in balance control.

## 7. Conclusions

From the many miscellaneous methods of modelling and controlling biped robots, we have attempted to collect the multi-level control strategies used for ZMP-based biped/humanoid robots and introduce them systematically. The designer should take into account the computational complexity of the target control schemes owed to the large DOFs of the target biped/humanoid robot. The efficacy of ASIMO humanoid robots may motivate researchers to investigate the control architecture of Fig. 10 in detail. It is clear that ASIMO designers depend on approximate models for walking generators and balance control combined with force/moment distributors and control. In fact, a thorough comparative study is needed to compare the familiar control architectures to determine the best control system for designing biped robots.

## References

[1]  P. Lima and M. I. Ribeiro," Mobile robotics," Course Handouts, Instituto Superior Técnico/Instituto de Sistemas e Robótica, Portugal, (March 2002).

[2]  M. F. Silva and J.T. Machado," A literature review on the optimization of legged robots," Journal of Vibration and Control 18(12), 1753-1767(2012).

[3]  G. A. Bekey," Autonomous Robots: From biological inspiration to implementation and control", MIT Press, USA (2005).

[4]  K. Hashimoto, Y. Sugahara, H. Sunazuka, C. Tanaka, A. Ohata, M. Kawase, H. –O. Lim and A. Takanishi" Biped landing pattern modification method with nonlinear compliance control," In IEEE International Conference on Robotics and Automation (ICRA 2006), Orlando, FL, pp. 1213-1218 (May 2006).

[5]  K. Hashimoto, A. Hayashi, T. Sawato, Y. Yoshimura, T. Asano, K. Hattori, Y. Sugahara, H.-O. Lim and A. Takanishi," Terrain-adaptive control to reduce landing impact force for human-carrying biped robot," In IEEE/ASME International Conference on Advanced Intelligent Mechatronics (AIM 2009), Singapore, pp. 174-179 (Jul. 2009).

[6]  K. Hashimoto, T. Sawato, A. Hayashi, Y. Yoshimura, T. Asano, K. Hattori, Y. Sugahara, H.-O Lim and A. Takanishi," Static and dynamic disturbance compensation control for a biped walking vehicle," In Proceedings of the 2nd Biennial IEEE RAS/EMBS International Conference on Biomedical Robotics and Biomechatronics, AZ, USA, pp. 457-462 (Oct. 2008).

[7]  K. Hashimoto, Y. Sugahara, C. Tanaka, A. Ohta, K. Hattori, T. Sawato, A. Hayashi, H.-O Lim and A. Takanishi," Unknown disturbance compensation control for a biped walking vehicle," In Proceedings of the 2007 IEEE/RSJ International Conference on Intelligent Robots and Systems, San Diego, USA, pp. 2204-2209 (2007).

[8]  Y. Sugahara, K. Hashimoto, N. Endo, T. Sawato, M. Kawase, A. Ohta, C. Tanaka, A. Hayashi, H.-O Lim and A. Takanishi," Development of a biped locomotor with the double stage linear actuator," In Proceedings of the 2007 IEEE International Conference on Robotics and Automation, Roma, Italy, pp.1850-1855(Apr. 2007).

[9]  K. Hashimoto, Y. Sugahara, A. Hayashi, M. Kawase, T. Sawato, N. Endo, A. Ohta, C. Tanaka, H.-O Lim, and A. Takanishi" New foot system adaptable to convex and concave surface," In Proceedings of the 2007 IEEE International Conference on Robotics and Automation, Roma, Italy, pp.1869-1874(2007).

[10]  Y. Sugahara, T. Endo, H.-O Lim and A. Takanishi," Design of a battery-powered multi-purpose bipedal locomotor with parallel mechanism," In Proceedings of the 2002 IEEE/RSJ International Conference on Intelligent Robots and Systems, Lausanne, Switzerland, pp. 2658-2663(Oct. 2002).

[11]  H.-O Lim, Y. Sugahara and A. Takanishi," Development of a biped locomotor applicable to medical and welfare fields," In Proceedings of the 2003 IEEE/ASME International Conference on Advanced Intelligent Mechatronics, pp. 950-955(2003).

[12]  C. L. Vaughan," Theories of bipedal walking: an odyssey," Journal of Biomechanics 36(4), 513-523(2003).

[13]  M. H. Raibert," Legged Robots that balance," MIT Press, USA (1986).

[14]  M. Vukobratovic and B. Borovac," Zero-moment point- thirty five years of its life," International Journal of Humanoid Robotics 1(1), 157-173(2004).

[15]  B. Vanderborght, R. Van Ham, B. Verrelst, M. Van Damme, D. Lefeber, "Overview of the Lucy project: dynamic stabilization of a biped powered by pneumatic artificial muscles." Advanced Robotics 22(10), 1027-1051 (2008).

[16]  C. L. Golliday and H. Hemami," An Approach to analyzing biped locomotion dynamics and designing robot locomotion controls," IEEE Transactions on Automatic Control AC-22(6), (1977).

[17]  D. Kim, S.-J. Seo and G.-T. Park," Zero-moment point trajectory modelling of a biped walking robot using an adaptive neuro-fuzzy system", IEE Proc. - Control Theory Appl. 152(4), 411-426 (2005).

[18]  C. Chevallereau, G. Bessonnet, G. Abba and Y. Aoustin," Bipedal Robots, Modeling , design  and building walking robots," John Wiley and Sons Inc, USA (2009).

[19]  M. Raibert, S.Tzafestas and  C. Tzafestas," Comparative simulation study of three control techniques applied to a biped robot," In Proc. International Conference on Systems, Man & Cybernetics System Engineering in the Service of Humans, Le Touquet, France, vol. 1, pp. 494-502 (1993).

[20]  W.-H Zhu,"Virtual decomposition control: towards hyper degrees of freedom," Springer–Verlag, Berlin (2010).

[21]  I.-W. Park, J.-Y. Kim and J.-H. Oh," Online biped walking pattern generation for humanoid robot KHR-3(KAIST Humanoid Robot-3: HUBO)," IEEE-RAS International Conference on Humanoid Robots,Genova,pp.398-403 (Dec. 2006).

[22]  J. Pratt," Exploiting inherent and natural dynamics in the control of bipedal walking robots," Ph.D. Dissertation, Massachusetts Institute of Technology (2000).

[23]  Hayder F. N. Al-Shuka, F. Allmendinger, B. Corves, W.-H. Zhu," Modeling, stability and walking pattern generators of biped robots: a review," Robotica, FirstView Article, pp. 1-28 (2013).

[24]  G. Ozyurt," 3-D Humanoid gait simulation using an optimal predictive control," MSc. Thesis, Middle East Technical University, Turkey (2005).

[25]  M. W. Whittle," Gait analysis: An Introduction," 4th Edition, Edinburgh, Butterworth-Heinemann, USA, 2007.

[26]  Q. Huang, S. Kajita, N. Koyachi, K. Kaneko, K. Yokoi, H. Arai, K. Komoriya and Kazuo Tane" A High stability, smooth walking pattern for a biped robot" In Proceedings of the 1999 IEEE International conference on Robotics and Automation, vol. 1, Detroit, MI, pp.65-71 (May 1999).

[27]  Q. Huang, K. Yokoi, S. Kajita, K. Kaneko, H. Arai, N. Koyachi and K.Tanie," Planning walking patterns for a biped robot," IEEE Transactions on Robotics and Automation 17(3), 280 - 289 (2001).

[28]  N. Handharu, J. Yoon and G. Kim," Gait pattern generation with knee stretch motion for biped robot using toe and heel joints," IEEE-RAS International Conference on Humanoid Robots, Daejeon, Korea, pp. 265-270 (Dec. 2008).

[29]  A. Goswami," Postural stability of biped robots and the foot-rotation indicator (FRI) point," The International Journal of Robotics Research 18(6), 523-533(1999).

[30]  Pieter van Zutven, D. Kostic and H. Nijmeijer," On the stability of bipedal walking," N. Ando et al. (Eds.): SIMPAR 2010, LNAI 6472, pp.521-532 (2010).

[31]  E. Nicholls," Bipedal dynamic walking in Robotics," Honors thesis, The University of Western Australia Department of Electrical and Electronic Engineering (1998).

[32]  J. Pratt, R. Tedrake," Velocity-based stability margins for fast bipedal walking," In Fast Motions in Biomechanics & Robotics, Lecture Notes in Control and Information Sciences ,vol. 340, pp.299-324 (2006).

[33]  J. Pratt, J. Carff, S. Drakunov and A. Goswami," Capture point: a step toward humanoid push recovery," IEEE-RAS International Conference on Humanoid Robot, Genova, pp. 200-207 (Dec 2006).

[34]  D.L. Wight, E.G. Kubica, D. W. L. Wang," Introduction to the foot placement estimator: a dynamic measure of balance for bipedal robotics", Journal of computational and nonlinear dynamics vol. 3, pp. 1-9(2008).

[35]  M. Vukobratovic and J. Stepanenko," On the stability of anthropomorphic systems," Mathematical Biosciences 15(1-2), 1–37 (1972).

[36]  S. Kajita and B. Espiau, ''legged robots,'' In: Springer Handbook of Robotics, B. Siciliano and O. Khatib, (Eds.), Springer, Berlin (2008).

[37]  A. G. Alba and T. Zielinska," Postural equilibrium criteria concerning feet properties for biped robots," Journal of Automation, mobile robotics and Intelligent Systems, 6(1), pp. 22-27, 2012.

[38]  C. Pop, A. Khajepour, J. P. Huissoon and A. E. Patla," Experimental/analytical analysis of human locomotion using bondgraphs," J Biomech Eng., 125(4):490-8 (2003).

[39]  D.G.E. Hobbelen and M. Wisse," Limit cycle walking", Humanoid Robots: Human-like machines, Matthias Hackel (Ed.), I-Tech, Vienna, Australia, pp. 277-294 (2007).

[40]  S.-H. Hyon, J. G. Hale, G. Cheng, "Full-body compliant human-humanoid interaction: balancing in the presence of unknown external forces," IEEE Transactions on Robotics 23 (5), 884-898 (Oct. 2007).

[41]  C. Ott, M. A. Roa, G. Hirzinger," Posture and balance control for biped robots based on contact force optimization," 11th IEEE-RAS International Conference on Humanoid Robots (Humanoids), Bled, Solvenia, pp. 26-33 (Oct. 2011).

[42]  J. Or and A.Takanishi,"A biologically inspired CPG-ZMP control system for the real-time balance of a single-legged belly dancing robot", IEEE/RSJ International Conference on Intelligent Robots and Systems, Sendai, Japan, vol. 1,931 - 936 (Oct. 2004).

[43]  J. H. Park and K.D. Kim," Biped robot walking using gravity-compensated inverted pendulum mode and computed torque control", In Proceedings of the 1998 IEEE International conference of Robotics & Automation, Leuven, Belgium,vol.4, pp. 3528-3533 (May 1998).

[44] Q. Huang and K. Ono," Energy-efficient walking for biped robot using self-excited mechanism and optimal trajectory planning," Humanoid Robots, New Developments, Armando Carlos de Pina Filho (Ed.), I-Tech, Vienna, Austria,pp.321-342 (2007).

[45] A. Takanishi, M. Ishida, Y. Yamazaki, and I. Kato," Realization of dynamic walking by the biped walking robot WL-10RD;" Int. Conf. Adv. Robot. (ICAR'85), pp. 459-466 (1985).

[46] K. Harada, S. Kajita, K. Kaneko and H. Hirukawa," Pushing manipulation by humanoid considering two-kinds of ZMPs," In Proceedings of the 2003 IEEE International Conference on Robotics and Automation, Taipei, Taiwan, vol.2, pp. 1627-1632 (Sept. 2003).

[47] P. Sardain and G. Bessonnet," Zero-moment point-Measurements from a human walker wearing robot feet as shoes," IEEE Transactions on systems, Man, and Cybernetics-Part A: Systems and Humans 34(5), 638-648 (2004).

[48] B. Yuksel, C. Zhou and and K. Leblebicioglu,"Ground reaction force analysis of biped locomotion," In Proceedings of the 2004 IEEE conference on Robotics, Automation and Mechatronics, Singapore, vol.1, pp. 330-335 (Dec 2004).

[49] A. Takanishi, T. Takeya, H. Karaki and I .Kato," A control method for dynamic biped walking under unknown external force," IEEE International Workshop on Intelligent Robots and Systems,IROS'90, Ibaraki, vol.2, pp. 795 - 801 (Jul. 1990).

[50] P. Sardain and G. Bessonnet," Force acting on a biped robot. Center of pressure-zero moment point," IEEE Transactions on systems, Man, and Cybernetics-Part A: Systems and Humans, 34(5), 630 - 637 (2004).

[51] T. Tsuji and K. Ohnishi, "A control of biped robot which applies inverted pendulum mode with virtual supporting point," In Proceedings of Advanced Motion Control, 7th Int. Workshop, Maribor, Solvenia, pp. 478-483(2002).

[52] N. Waki, K. Matsumoto and A. Kawamura," Lateral sway motion generation for biped robots using virtual supporting point," The 11th IEEE Int. workshop on Advanced Motion Control, Nagaoka, Japan, pp. 124-128 (March 2010).

[53] H. Herr and M. Popovic," Angular momentum in human walking," The Journal of Experimental Biology, vol. 211, pp. 467-481 (2008).

[54] M. B. Popovic and H. Herr, "Ground reference points in legged locomotion: Definitions, biological trajectories and control implications," In: Mobile robots towards new applications, A. Lazinica (ed.), Germany, pp. 79-104 (2006).

[55] A. Goswami and V. Kallem," Rate of change of angular momentum and balance maintenance of biped robots," In Proceedings of the 2004 IEEE International Conference on Robotics and Automation, New Orleans, LA, vol. 4, pp. 3785-3790 (2004).

[56] A. Sano and J. Furusho," Realization of natural dynamic walking using the angular momentum information," IEEE International Conference on Robotics and Automation, Cincinnati, OH, vol. 3, pp. 1476-1481 (May 1990).

[57] A. Sano and J. Furusho," Control of torque distribution for the BLR-G2 biped robot," 5th International Conference on Advanced Robotics, Pisa, Italy, vol. 1, pp.729 – 734 (June 1991).

[58] M. Popovic, A. Hofmann and H. Herr," Angular momentum regulation during human walking: biomechanics and control," IEEE International Conference on Robotics and Automation, New Orleans, LA, USA, vol. 3, pp. 2405-2411 (2004).

[59] M. Popovic, A. Englehart and H. Herr," Angular momentum primitives for human walking: biomechanics and control," In Proceedings of te 2004 IEEE/RSJ International Conference on Intelligent Robots and Systems, Sendai, Japan, vol. 2, pp. 1685-1691 (2004).

[60] M. Popovic, A. Hofmann, H. Herr," Zero spin angular momentum control: Definition and applicability," IEEE-RAS/RSJ International Conference on Humanoid Robots, Los Angeles, CA, Vol. 1, pp. 478 - 493 (2004).

[61] M. Popovic and H. Herr," Global motion control and support base planning," IEEE/RSJ International Conference on Intelligent Robots and Systems, Alberta, Canada, pp. 3877 - 3884 (2005).

[62] S. Kajita, F. Kanehiro, K. Kaneko, K. Fujiwara, K. Harada, K. Yokoi and H. Hirukawa," Resolved momentum control: humanoid motion planning based on the linear and angular momentum," In Proceedings of the 2003 IEEE/RSJ International Conference on Intelligent Robots and Systems (IROS), vol. 2, pp. 1644-1650 (2003).

[63] N. E. Sian, K. Yokoi, S. Kajita, F. Kanchiro and K. Tanie," Whole body teleoperation of a humanoid robot-A method of integrating operator's intention and robot's autonomy-," In Proceedings of the 2003 IEEE/RSJ International Conference on Intelligent Robots and Systems (IROS), Taipei, Taiwan, vol. 2, pp. 1613-1619 (Sept. 2003).

[64] J. Pratt and G. Pratt," Intuitive control of a planar bipedal walking robot," In Proceedings of the IEEE International Conference on Robotics and Automation (ICRA), vol. 3, pp. 2014-2021(1998).

[65] H. F. N. Al-Shuka, B. Corves, B. Vanderborght and W.-H. Zhu,'' Finite difference-based suboptimal walking pattern generators of biped robot with continuous dynamic response," International Journal of Modeling and Optimization, 3(4), pp. 337-343 (2013).

[66] J. Rebula, I. Canas, J. Pratt and A. Goswami," Learning capture points for humanoid push recovery," IEEE/RAS International Conference on Humanoid Robots, Pittsburgh, PA, pp. 65-72 (Nov. 2007).

[67] C. Azevedo, B. Espiau, B. Amblard, C. Assaiante," Bipedal locomotion: toward unified concepts in robotics and neuroscience," Biol Cybern 96(2): 209-228 (2007).

[68] F. E. Huxham, P. A. Goldie and A. E. Patla," Theoretical considerations in balance assessment," Australian Journal of Physiotherapy 47 (2), pp. 89-100 (2001).

[69] G. Bessonnet, S. Chesse and P. Sardain," Generating optimal gait of a human-sized biped robot," In the 5th Int. Conf. Climbing and Walking Robots, Paris, pp.241-253(2002).

[70] H. F. N. Al-Shuka, B. Corves and W-H Zhu," On the dynamic optimization of biped robot," Lecture of Notes on Software Engineering, vol.1 (3), pp. 237-243 (2013).

[71] S. Kajita and K. Tani," Experimental study of biped dynamic walking in the linear inverted pendulum mode," In Proc. IEEE Conf. Robotics and Automation, vol.3, pp. 2885-2891 (1995).

[72] Z. Tang, C. Zhou, Z. Sun,'' Trajectory planning for smooth transition of a biped robot,'' In Proceedings of the IEEE International Conference on Robotics and Automations, vol. 2, pp. 2455-2460 (2003).

[73] L. Wang, Z. Yu, F. He, Y. Jiao,'' Research on biped robot gait in double-support phase,'' IEEE International Conference on Mechatronics and Automation, pp. 1553-1558 (2007).

[74] C.-L. Shih,''Gait synthesis for a biped robot,'' Robotica, 15 (6), pp. 599-607(1997).

[75] X. Mu and Q. Wu," Synthesis of a complete sagittal gait cycle for a five-link biped robot,'' Robotica, 21(5), pp. 581-587(2003).

[76] A. Takanishi, H.-O Lim, M. Tsuda, I. Kato,'' Realization of dynamic biped walking stabilized by trunk motion on a sagittally uneven surface, '' IEEE International Workshop on Intelligent Robots and Systems, vol.1 pp. 323-330 (1990).

[77] K. Harada, S. Kajita, F. Kanehiro, K. Fujiwara, K. Kaneko, K. Yokoi and H. Hirukawa" Real-time planning of humanoid robot's gait for force-controlled manipulation," IEEE/ASME Transactions on Mechatronics 12(1): 53-62 (2007).

[78] I.-W. Park, J.-Y. Kim and J.-H. Oh, "Online walking pattern generation and its application to a biped humanoid robot-KHR-3 (HUBO)," Advanced Robotics 22(2-3), 159-190 (2008).

[79] Y. Kuroki, B. Blank, T. Mikami, P. Mayeux, A. Miyamoto, R. Playter, K. Nagasaka, M. Raibert, M. Nagano, J. Yamaguchi," Motion creating system for a small biped entertainment robot," In Proceedings of the IEEE/RSJ International conference on Intelligent Robots and Systems (IROS), Las Vegas, Nevada, vol.2 , pp. 1394-1399 (Oct. 2003).

[80] J. Kuffner, K. Nishiwaki, S. Kagami, Y. Kuniyoshi, M. Inaba, H. Inoue," Self-collision detection and prevention for humanoid robots," In Proceedings of the IEEE International Conference on Robotics and Automation (ICRA), Washington, DC, vol. 3, pp. 2265-2270 (May 2002).

[81] B. Mirtich," V-Clip: Fast and robust polyhedral collision detection," ACM Transactions on Graphics (TOG) 17(3): 177-208(1998).

[82] H. Sugiura, M. Gienger, H. Janssen, C. Goerick," Real-time collision avoidance with whole body motion control for humanoid robots," In Proceedings of the IEEE/RSJ International Conference on Intelligent Robots and Systems, San Diego, CA, pp. 2053-2058 (Oct. 2007).

[83] O. Stasse, A. Escande, N. Mansard, S. Miossec, P. Ervrard and A. Kheddar," Real-time (self)-collision avoidance task on a HRP-2 humanoid robot," IEEE International Conference on Robotics and Automation (ICRA), Pasadena, CA, USA, pp. 3200-3205 (May 2008).

[84] A. Safonova N. Pollard and J. K. Hodgins," Optimizing human motion for the control of a humanoid robot," 2nd International Symposium on Adaptive Motion and Animals and Machines (AMAM) (March 2003).

[85] J. Kuffner, S. Kagami, K. Nishiwaki, M. Inaba and H. Inoue" Online footstep planning for humanoid robots," In Proceedings of the IEEE International Conference on Robotics and Automation, Taipei, Taiwan, vol. 1, pp. 932-937 (Sept. 2003).

[86] B. Faverjon, P. Tournassoud," A local based approach for path planning of manipulators with a high number of degrees of freedom," IEEE International Conference on Robotics and automation, vol. 4, pp. 1152-1159 (March 1987).

[87] F. Kanehiro, Lamiraux, O. Kanoun, E. Yoshida and J.-P. Laumond," A local collision avoidance method for non-strictly convex polyhedra," In Proceedings of: Science and Systems, pp. 151-158 (2009).

[88] A. Escande, S. Miossec and A. Kheddar," Continuous gradient proximity distance for humanoids free-collision optimized-postures," 7th IEEE-RAS International Conference on Humanoid Robots, Pittsburgh, PA, pp. 188-195 (2007).

[89] S. Nakaoka, A. Nakazawa, F. Kanehiro, K. Kaneko, M. Morisawa, K. Ikeuchi," Task model of lower body motion for a biped humanoid robot to imitate human dances," IEEE/RSJ International Conference on Intelligent Robots and Systems, pp. 3157-3162 (Aug. 2005).

[90] S. Kajita and K. Tani," Experimental study of biped dynamic walking," IEEE, Control Systems 16(1):13-19 (1996).

[91] F. Miyazaki and S. Arimoto," A control theoretic study on dynamical biped locomotion," Journal of Dynamic Systems, Measurement, and Control, vol. 102, pp. 233-239 (1980).

[92] R. C. Cannon," Dynamics of physical systems," New York, McGraw- Hill (1967).

[93] J. Schaefer," On the bounded control of some unstable mechanical systems," Dept. of Aeronautics and Astronautics, Stanford University, SUDAR Rep. 233 (Apr. 1965).

[94] D. C. Witt," A feasibility study of powered-limb prosthesis," In Proceedings of the Institution of Mechanical Engineers, 183(10),18-25 (1968).

[95] H. Hemami, F. C. Weimer and S. H. Koozekanani," Some aspects of the inverted pendulum problem for modeling of locomotion systems," IEEE Transactions on Automatic control, 18 (6), 658-661(1973).

[96] F. Gubina, H. Hemami and R. B. McGhee," On the dynamic stability of biped locomotion," IEEE Transactions on Biomedical Engineering, BME-21(2), 102-108 (1974).

[97] H. Miura & I. Shimoyama," Dynamic walk of a biped" International Journal of Robotics Research 3(2), 60-74 (1984).

[98] S. Kajita, T. Yamaura and A. Kobayashi," Dynamic walking control of a biped robot along a potential energy conserving orbit", IEEE Transaction on Robotics and Automation 8(4), 431-438 (1992).

[99] J. E. Pratt and S. V. Drakunov," Derivation and application of a conserved orbital energy for the inverted pendulum bipedal walking model," IEEE International Conference on Robotics and Automation, Roma, pp. 4653-4660(2007).

[100]M. Shibuya, T. Suzuki and K. Ohnishi," Trajectory planning of biped robot using linear pendulum mode for double support phase," in Proc. IECON 2006-32nd Annual Conf. IEEE Industrial Electronics, pp.4094-4099 (2006).

[101]S. Kudoh and T. Komura," $C^2$ continuous gait-pattern generation for biped robots," in Proc. 2003 IEEE/RSJ Intelligent Robots and Systems, Las Vegas, Nevada, vol.2, pp. 1135-1140 (2003).

[102]H. F. N. Al-Shuka, B. Corves, Wen-Hong Zhu, Bram Vanderborght," A simple algorithm for generating stable biped walking patterns " International Journal of Intelligent Computer Applications, 101 (4):29-33 (2014).

[103]M. –S. Kim, I. Kim, S. Park, J. H. Oh," Realization of stretch-legged walking of the humanoid robot," 8th IEEE-RAS International Conference on Humanoid Robots, Daejeon, Korea, pp. 118-124 (Dec. 2008).

[104]Y. Ogura, T. Kataoka, H. Aikawa, K. Shimomura, H. –O. Lim and A. Takanishi," Evaluation of various walking patterns of biped humanoid robot," In Proceedings of the IEEE International Conference on Robotics and Automation , Barcelona, Spain, pp. 603-608 (Apr. 2005).

[105]R. Kurazume, S. Tanaka, M. Yamashita, T. Hasegawa and K. Yoneda," Straight legged walking of a biped robot," IEEE/RSJ International Conference on Intelligent Robots and Systems, pp. 3095-3101(2005).

[106]A. Albert and W. Gerth, "Analytic path planning algorithms for bipedal robots without a trunk," Journal of Intelligent and Robotic Systems 36 (2),109-127(2003).

[107]A. Takanishi, M. Tochizawa, H.Karaki and I. Kato," Dynamic biped walking stabilized with optimal trunk and waist motion", IEEE/RSJ International Workshop on Intelligent Robots and Systems, Tsukuba, Japan, pp. 187–192 (Sept.1989).

[108]T. Ha and C.-H. Choi, "An effective trajectory generation method for bipedal walking," Robotics and Autonomous Systems 55(10), 795–810(2007).

[109]T. Sugihara, Y. Nakamura and H. Inoue," Realtime humanoid motion generation through ZMP manipulation based on inverted pendulum control," In Proceedings of the IEEE International Conference on Robotics and Automation, Washington, DC, vol. 2, pp. 1404-1409 (May 2002).

[110]Y. Choi, B.-J. You and S.-R. Oh," On the Stability of Indirect ZMP Controller for Biped Robot Systems," IEEE/RSJ International Conference on Intelligent Robots and Systems, Sendai, Japan, vol. 2, pp. 1966-1971 (Oct. 2004).

[111]Napoleon, S. Nakaura, M. Sampei," Balance control analysis of humanoid robot based on ZMP feedback control," In the proceedings of the IEEE/RSJ International Conference on Intelligent Robots and Systems, vol. 3, pp. 2437-2442 (2002).

[112]S. Hong, Y. Oh, Y.-H. Chang and B.-J. You," Walking pattern generation for Humanoid robots with LQR and feedforward control method," 34th Annual Conference of IEEE on Industrial Electronics, Orlando, FL, pp. 1698-1703 (Nov. 2008).

[113]S. Kajita, F. Kanehiro, K. Kaneko, K. Fujiwara, K. Harada, K. Yokoi and H. Hirukawa," Biped walking pattern generation by using preview control of zero-moment point," In Proceedings of the IEEE International Conference on Robotics and Automation, vol. 2, pp. 1620-1626 (Sept. 2003).

[114]T. Takenaka, T. Matsumoto and T. Yoshiike," Real time motion generation and control for biped robot-1st Report: walking gait pattern generation," IEEE/RSJ International Conference on Intelligent Robots and Systems, St. Louis, MO, pp. 1084-1091 (Oct. 2009).

[115]S. Kajita, M. Morisawa, K. Harada, K. Kaneko, F. Kanehiro, K. Fujiwara and H. Hirukawa," Biped walking pattern generator allowing auxiliary ZMP control," In Proceedings of the IEEE/RSJ International Conference on Intelligent Robots and Systems, Beijing, China, pp. 2993-2999 (Oct. 2006).

[116]S. Shimmyo and K. Ohnishi," Nested preview control by utilizing virtual plane for biped walking pattern generation including COG up-down motion," 36th Annual Conference on IEEE Industrial Electronics Society , Glendale, AZ, pp. 1571-1576 (Nov. 2010).

[117] S. Czarnetzki, S. Korner and O. Urbann," Observer-based dynamic walking control for biped robots," Robotics and Autonomous Systems 57(8): 839-845 (2009).

[118]P.-B. Wieber," Trajectory free linear model predictive control for stable walking in the presence of strong perturbations," 6th IEEE-RAS International Conference on Humanoid Robots, Genova, pp. 137-142 (Dec. 2006).

[119] D. Dimitrov, P.-B. Wieber, H.J. Ferreau and M. Diehl," On the implementation of model predictive control for on-line walking pattern generation," IEEE International Conference on Robotics and Automation (ICRA), Pasadena, CA, pp. 2685-2690 (May 2008).

[120]S. Wright," Applying new optimization algorithms to model predictive control," 5th International Conference on Chemical Process Control-CPC-V (1996).

[121]R. A. Bartlett, A. Wächter and L. T. Biegler," Active set vs. interior point strategies for model predictive control," In the Proceedings of the American Control Conference, Chicago, IL, vol. 6, pp. 4229-4233 (2000).

[122]M. Arbulu, D. Kaynov and C. Balaguer," The RH-1 full-size humanoid robot: control system design and walking pattern generation," In Climbing and Walking Robots edited by B. Miripour, InTech (2010).

[123]B.-J. Lee, D. Stonier, Y.-D. Kim, J.-K. Yoo, J. –K. Yoo and J.-H. Kim," Modifiable walking pattern of a humanoid robot by using allowable ZMP variation," IEEE Transactions on Robotics 24(4): 917-925 (2008).

[124]S. Hong, Y. Oh, Y.-H. Chang, B.-J. You," A walking pattern generation method for humanoid robots using least square method and quartic polynomial," Humnoid Robots, Ben Choi (Ed.), ISBN: 978-953-7619-44-2, InTech (2009).

[125]K. Harada, S. Kajita, K. Kaneko and H. Hirukawa," An analytical method on real-time gait planning for a humanoid robot ," IEEE/RAS International Conference on Humanoid Robots, vol. 2, pp. 640-655 (2004).

[126]M. Morisawa, K. Harada, S. Kajita, K. Kaneko, F. Kanehiro, K. Fujiwara, S. Nakaoka and H. Hirukawa," A biped pattern generation allowing immediate modification of foot placement in real-time," 6th IEEE-RAS International Conference on Humanoid Robots, pp. 581-586 (2006).

[127]M. Morisawa, K. Harada, S. Kajita, K. Kaneko, J. Sola, E. Yoshida, N. Mansard, K.Yokoi and J.-P. Laumond," Reactive stepping to prevent falling for humanoids," 9th IEEE-RAS International Conference on Humanoid Robots, Paris, France, pp. 528-534 (2009).

[128]M. Morisawa, F. Kanehiro, K. Kaneko, N. Mansard, J. Sola, E. Yoshida, K. Yokoi and J.-P. Laumond," Combining suppression of the disturbance and reactive stepping for recovering balance," IEEE/RSJ International Conference on Intelligent Robots and Systems (IROS), pp. 3150-3156(2010).

[129]R. Tajima, D. Honda, K. Suga, "Fast Running experiments involving a humanoid robot," IEEE Int. conf. on Robotics and Automation, Kobe, Japan, pp. 1571-1576 (May 2009).

[130]T. Komura, H. Leung, S. Kudoh and J. Kuffner," A feedback controller for biped humanoids that can counteract large perturbations during gait," In Proceedings of the IEEE International Conference on Robotics and Automation, Barcelona, Spain, pp. 1989-1995(Apr. 2005).

[131]F. B. Horak and L. M. Nasher," Central programming of postural movements: adaptation to altered support-surface configurations," Journal of Neurophysiology 55(6), 1369-1381 (1986).

[132]A. D. Kuo," An optimal control model for analyzing human postural balance," IEEE Transactions on Biomedical Engineering 42(1), 87-101 (1995).

[133]A. V. Alexandrov, A. A. Frolov and J. Massion," Biomechanical analysis of movement strategies in human forward trunk bending. I. Modeling," Biol Cybern 84(6): 425-434 (2001).

[134]C. F. Runge, C. L. Shupert, F. B. Horak and F. E. Zajac" Ankle and hip postural strategies defined by joint torques,'' Gait and Posture 10(2),161-170(1999).

[135]D.N. Nenchev and A. Nishio," Experimental validation of ankle and hip strategies for balance recovery with a biped subjected to an impact," In Proceedings of the IEEE/RSJ International Conference on Robots and Systems, San Diego, CA, pp. 4035-4040 (Nov. 2007).

[136]A. Nishio, K. Takahashi and D. N. Nenchev," Balance control of a humanoid robot based on the reaction null space method," In Proceedings of the IEEE/RSJ International Conference on Intelligent Robots and Systems, pp. 1996-2001 (2006).

[137]H. Ono, T. Sato and K. Ohnishi," Balance recovery of ankle strategy: using knee joint for biped robot," 1st International Symposium on Access Space (ISAS), Yokohama, pp. 236-241 (2011).

[138]C.-S. Park, T. Ha, J. Kim, C.-H. Choi," Trajectory generation and control for a biped robot walking upstairs," International Journal of Control, Automation, and Systems 8(2):339-335 (2010).

[139]V. Prahald, G. Dip and C. M. –Hwee," Disturbance rejection by online ZMP compensation,'' Robotica 26(1):9-17 (2008).

[140]S. Kajita, K. Yokoi, M. Saigo and K. Tanie, "Balancing a humanoid robot using backdrive concerned torque control and direct angular momentum feedback," In Proceedings of the IEEE International Conference on Robotics and Automation, Seoul, Korea, vol. 4, pp. 3376-3382 (May 2001).

[141]N. Oda and M. Ito," Experimental study of walking motion stabilization for biped robot with flexible ankle joints," 35th Annual Conference of IEEE on Industrial Electronics (IECON'09), pp. 4197-4202 (Nov. 2009).

[142]T. Sato and K. Ohnishi," ZMP disturbance observer for walking stabilization of biped robot," 10th IEEE International Workshop on Advanced Motion Control (AMC'08), Trento, pp. 290-295(March 2008).

[143]T. Takenaka, T. Matsumoto and T. Yoshiike," Real time motion generation and control for biped robot-3rd Report: dynamics error compensation," IEEE/RSJ International Conference on Intelligent Robots and Systems, St. Louis, MO, pp. 1594-1600 (Oct. 2009).

[144]T. Takenaka, T. Matsumoto, T. Yoshiike, T. Hasegawa, S. Shirokura, H. Kaneko and A. Orita," Real time motion generation and control for biped robot-4th Report: Integrated balance control-," IEEE/RSJ International Conference on Intelligent robots and Systems, St. Louis, USA, pp. 1601-1608(Oct. 2009).

[145]M. Sobotka, D. Wollherr and M. Buss," A Jacobian method for online modification of precalculated  gait trajectories," In Proceedings of the 6th International Conference on Climbing and Walking Robots (CLAWAR 2003), Catania, Italy, pp. 435-442 (2003).

[146]D. Wollherr and M. Buss," Posture modification for biped humanoid robots based on Jacobian method," In Proceedings of the IEEE/RSJ International Conference on Intelligent Robots and Systems, Sendai, Japan, vol. 1, pp. 124-129 (Oct. 2004).

[147]O. Brock, O. Khatib and S. Viji," Task-consistent obstacle avoidance and motion behavior for mobile manipulation," In Proceedings of the IEEE International Conference on Robotics and Automation (ICRA'02), Washington, DC, vol. 1, pp. 388-393 (May 2002).

[148]K. Yamane and Y. Nakamur," Dynamics filter-concept and implementation of online motion generator for human figures," IEEE Transactions on Robotics an Automation 19 (3), 421-432 (2003).

[149]S. Tak, O. –Y. song and H. –S. Ko, "Motion balance filtering," Computer Graphics Forum 19 (3): 437-446 (2000).

[150]Y. Nakamura and K. Yamane," Dynamics computations of structure-varying kinematic chains and its application to human figures," IEEE Transactions on Robotics and automation 16 (2), 124-134 (Apr. 2000).

[151]P. M. Silva and J. A. T. Machado," Towards force interaction control of biped walking robots," In Proceedings of the IEEE/RSJ International Conference on Intelligent Robots and Systems (IROS 2004), Sendai, Japan, vol. 3, pp. 2568-2573 (Oct. 2004).

[152]S. Setiawan, S. Hyon, J. Yamaguchi and A. Takanishi," Physical interaction between human and a bipedal humanoid robot – realization of human-follow walking," In Proceedings of the IEEE International Conference on Robotics and Automation, Detroit, MI, vol. 1, pp. 361-367 (May 1999).

[153]S.-H. Hyon, R. Osu and Y. Otaka," Integration of multi-level postural balancing on humanoid robots," IEEE International Conference on Robotics and Automation, Kobe International Conference Center, Kobe, Japan, pp.1549- 1556 (May 2009).

[154]S.-H. Hyon and G. Cheng," Disturbance rejection for biped humanoids," IEEE International Conference on Robotics and Automation, Roma, Italy, pp. 2668-2675 (Apr. 2007).

[155]S.-H. Hyon and G. Cheng," Passivity-based full-body force control for humanoids and application to dynamic balancing and locomotion," In Proceedings of the IEEE/RSJ International Conference on Intelligent Robots and Systems, Beijing, China, pp. 4915-4922 (Oct. 2006).

[156]C. Azevedo, P. Poignet and B. Espiau," Moving horizon control for biped robots without reference trajectory," In Proceedings of the IEEE International Conference on Robotics and Automation, Washington, DC, vol. 3, pp. 2762-2767 (May 2002).

[157]E. F. Camacho and C. Bordons," Model predictive control," Springer-Verlag, London, 2007.

[158]P.-B. Wieber," Viability and predictive control for safe locomotion," IEEE/RSJ International Conference on Robots and Systems, Nice, France, pp. 1103-1108 (Sept. 2008).

[159]H. Diedam, D. Dimitrov, P.-B. Wieber, K. Mombaur and M. Diehl," Online walking gait generation with adaptive foot positioning through linear model predictive control," IEEE/RSJ International Conference on Intelligent Robots and Systems, Nice, France, pp. 1121-1126 (Sept. 2008).

[160]C. Azevedo, P. Poignet and B. Espiau," Artificial locomotion control from human to robots," Robotics and Autonomous Systems 47 (4): 203-223 (2004).

[161]Y. Yin and S. Hosoe," Mixed logic dynamical modeling and on line optimal control of biped robot," In: Humanoid Robots: Human-like Motion, Book Edited by Mathias Hackel, Vienna, Austria, pp. 315-328 (June 2007).

[162]F. Lydoire and P. Poignet," Nonlinear predictive control using constraint satisfaction," 2nd International Workshop on Global Optimization and Constraint Satisfaction (COCOS), Lausanne, Switzerland (Nov. 2003).

[163]M. Parsa and M. Farrokhi," Robust nonlinear model predictive trajectory free control of biped robots based on nonlinear disturbance observer," 18th Iranian Conference on Electrical Engineering (ICEE), Isfahan, Iran, pp. 617-622 (May 2010).

[164]N. Kalamian and M. Farrokhi," Stepping of biped robots over large obstacles using NMPC controller," 2nd International Conference on Control, Instrumentation and Automation (ICCIA), Shiraz, Iran, pp. 917-922 (2011).

[165]D. Katic and M. Vukobratovic," Survey of intelligent control techniques for humanoid robots," Journal of Intelligent and Robotic Systems 37 (2): 117-141 (2003).

[166]J. Hill and F. Fahimi," Active disturbance rejection for walking bipedal robots using the acceleration of the upper limbs," Robotica, FirstView Article, pp. 1-18 (Mar. 2014).

[167] P. -B. Wieber," Holonomy and nonholonomy in the dynamics of articulated motion," In Fast motions in biomechanics and robotics, Lecture Notes in Control and Information Science, M. Diehl and K. Mombaur (Eds.), vol. 340, pp. 411-425 (2006).

[168] M. W. Spong and M. Vidyasagar," Robot dynamics and control," John Wiley and Sons, USA (1989).

[169] F. L. Lewis, D. M. Dawson and C. T. Abdallah," Robot manipulator control: Theory and practice," Marcel Dekker, Inc., New York, USA(2006).

[170] R. Kelly, V. Santibanez and A. Loria," Control of robot manipulators in joint space," Springer-Verlag London Limited (2005).

[171] W. –H. Zhu," Dynamics of general constrained robots derived from rigid bodies," J. Applied Mechanics, ASME, 75(3), 1-11 (2008).

[172] J. H. Park and H. Chung," Hybrid control for biped robots using impedance control and computed-torque control," In Proceedings of the IEEE International Conference on Robotics and Automation, Detroit, MI, vol. 2, pp. 1365-1370 (1999).

[173] X. Mu and Q. Wu," Development of a complete dynamic model of a planar five-link biped and sliding mode control of its locomotion during the double support phase," International Journal of Control 77(8): 789-799 (2004).

[174] S. A. A. Moosavian, A. Takhmar and M. Alghooneh," Regulating sliding mode control of a biped robot," In Proceedings of the International Conference on Mechatronics and Automation (ICMA 2007), Harbin, China, pp. 1547-1552 (Aug. 2007).

[175] A. B. Pournazhdi, M. Mirzaei, A. R. Ghiasi," Dynamic modeling and sliding mode control for fast walking of seven-link biped robot," 2$^{nd}$ International Conference on Control, Instrumentation, and Automation (ICCIA), Shiraz, Iran, pp. 1012-1017 (Dec. 2011).

[176] R. C. Luo, C.-W. Tzeng, P.-Z. Cheng and K.-W. Lee," Trajectory- tracking of nonlinear biped robot system based on adaptive fuzzy sliding mode control," The 33$^{rd}$ Annual Conference of the IEEE on Industrial Electronics Society (IECON 2007), Tapei, Taiwan, pp. 2789-2794 (Nov. 2007).

[177] C.-L. Hwang," A trajectory tracking of biped robots using fuzzy-model-based sliding-mode control," In Proceedings of the 41$^{st}$ IEEE Conference on Decision and Control, vol. 1, pp. 203-208(Dec. 2002).

[178] S. H. Lee, J. B. Park, Y. H. Choi," Sliding mode control based on self-recurrent wavelet neural network for five-link biped robot," International Joint Conference (SICE-ICASE), Busan, Korea, pp. 726-731 (Oct. 2006).

[179] G. Bartolini, G. Casalino, M. Aicardi," Learning and variable structure techniques in the control of a mechanical byped," 28$^{th}$ IEEE Conference on Decision and Control, Tampa, FL, vol. 3, pp. 2621-2628 (Dec. 1989).

[180] J. Kho and D. Lim," A Learning controller for repetitive gait control of biped walking robot," SICE 2004 Annual Conference, Sapporo, vol. 1, pp. 885-889 (Aug. 2004).

[181] L. Wang, Z. Liu, C. L. Chen, Y. Zhang, S. Lee, X. Chen," Energy-efficient SVM learning control system for biped walking robots," IEEE Transactions on Neural Networks and Learning Systems 24(5):831-837 (2013).

[182] Hayder F. N. Al-Shuka, B. Corves and W.-H. Zhu," Function approximation technique-based adaptive virtual decomposition control for a serial chain manipulator," Robotica, FirstView Article, pp.1-25 (2013).

[183] H.-O. Lim, S. A. Setiawan and A. Takanishi," Balance and impedance control for biped humanoid robot locomotion," In Proceedings of the IEEE/RSJ International Conference on Intelligent Robots and Systems, Maui, HI, vol. 1, pp. 494-499 (2001).

[184] J. H. Park and H. Chung," Impedance control and modulation for stable footing in locomotion of biped robots," In Proceedings of the IEEE/RSJ International Conference on Intelligent Robots and Systems (IROS'99), Kyongju, vol. 3, pp. 1786-1791 (1999).

[185] J. H. Park," Impedance control for biped robot locomotion," IEEE Transactions on Robotics and Automation 17(6), 870-82 (2000).

[186] K. Dupree, C.-H. Liang, G. Hu and W. E. Dixon," Adaptive Lyapunov-based control of a robot and mass-spring system undergoing an impact collision," IEEE Transactions on Systems, Man and Cybernetics, 38(4),1050-1061 (Aug. 2008).

[187] A. Tornambe," Modeling and control of impact in mechanical systems: theory and experimental results," IEEE Transactions on Automatic Control, 44(2), 294-309 (Feb. 1999).

[188] M. Machado, P. Moreira, P. Flores and H. M. Lankarani," Compliant contact force models in multibody dynamics: Evolution of the Hertz contact theory," Mechanism and Machine Theory, 53, pp. 99-121 (2012).

[189] C. Rengifo, Y. Aoustin, C. Chevallereau and F. Plestan," A penalty-based approach for contact forces computation in bipedal robots," 9$^{th}$ IEEE-RAS International Conference on Humanoid Robots, Paris, France, p. 121-127 (2009).

[190] M. Vukobratovic, V. Potkonjak and V. Matijevic," Dynamics of robots with contact tasks," Kluwer Academic Publishers, The Netherlands (2003).

[191] O. Bruneau and F. B. Ouezdou," Compliant contact of walking robot feet," In Proc. of the 3$^{rd}$ ECDP International Conference on Advanced Robotics, Intelligent Automation and Active Systems, Bremen, Germany (Sept. 1997).

[192] Y. Hwang, E. Inohira, A. Konno and M. Uchiyama," An order n dynamic simulator for a humanoid robot with a virtual spring-damper contact model," In Proceedings of the IEEE International Conference on Robotics and Automation (ICRA'03), vol. 1, pp. 31-36 (Sept. 2003).

[193] T. Buschmann, S. Lohmeier, H. Ulbrich and F. Pfeiffer," Dynamics simulation for a biped robot: modeling and experimental verification," In Proceedings of the IEEE International Conference on Robotics and Automation (ICRA'2006), Orlando, FL, pp. 2673-2678 (May 2006).

[194] H. Hirukawa, F. Kanehiro, S. Kajita, K. Fujiwara, K. Yokoi, K. Kaneko and K. Harada, " Experimental evaluation of the dynamic simulation of biped walking of humanoid robots," In Proceedings of the IEEE International Conference on Robotics and Automation (ICRA'03), vol. 2, pp. 1640-1645 (Sept. 2003).

[195] W. Blajer, W. O. Schiehlen," A control scheme for biped walking without impacts," Lecture Notes in Control and Information Sciences, vol. 187, pp. 311-321 (1993), .

[196] W. Blajer, W. Schiehlen," Walking without impacts as a motion /force control problem," J. Dyn. Meas. Control 114(4):660-665 (Dec. 1992).

[197] P. Seguin, G. Bessonnet," Generating optimal walking cycles using spline-based state-parameterization," Int. J. Human. Robot. 2(1): 47-80 (2005).

[198] M. Shibata, T. Natori," Impact force reduction for biped robot based on decoupling COG control scheme," 6$^{th}$ International Workshop on Advanced Motion Control, Nagoya, Japan, pp. 612-617 (Apr. 2000).

[199] Z. Liu, Y. Zhang, Y. Wang," A Type-2 fuzzy switching control system for biped robots," IEEE Transactions on Systems, Man and Cybernetcis, 37 (6), 1202-1213 (Nov. 2007).

[200] B. Koopman, H. J. Grootenboer, H. J. Jongh," An inverse dynamics model for the analysis, reconstruction and prediction of bipedal walking," Journal of Biomechanics, 28 (11), 1369-1376 (1995).

[201] L. Ren, R. K. Johnes, D. Howard," Whole body inverse dynamics over a complete gait cycle based only on measured kinematics," Journal of Biomechanics, 41(12), 2750-2759(2008).

[202] H. Dallali, G. A. Medrano-Corda and M. Brown," A comparison of multivariable and decentralized control strategies for robust humanoid walking," In UKACC International Conference on Control, Coventry, UK. (Sept. 2010).

[203] W. Blajer, D. Bestle, W. Schiehlen, An orthogonal complement matrix formulation for constrained multibody systems, Journal of Mechanical Design, vol. 116 (1994).

[204] E. Pennestri and P. P. Valentini, Coordinate reduction strategies in multibody dynamics: a review, In Atti Conference on Multibody System Dynamics, 2007.

[205] G. Zeng and A. Hemami, An overview of robot force control," Robotica, 15(5), 1997, 473-482.

[206] Hayder F. N. Al-Shuka, B. Corves, Wen-Hong Zhu," Dynamic modeling of biped robot using Lagrangian and recursive Newton-Euler formulations," International Journal of Computer Applications, 101(3):1-8 (2014).

[207] M. Vukobratovic and D. Juricic," Contribution to the synthesis of biped gait, IEEE Trans. Bio-Medical Eng. 16(1) 1969.

[208] K. Y. Yi," Locomotion of a biped robot with compliant ankle joints," In Proceedings of the IEEE International Conference on Robotics and Automation, Albuquerque, NM, vol. 1, pp. 199-204 (Apr. 1997).

# YOUR KNOWLEDGE HAS VALUE

- We will publish your bachelor's and
  master's thesis, essays and papers

- Your own eBook and book -
  sold worldwide in all relevant shops

- Earn money with each sale

Upload your text at www.GRIN.com
and publish for free